Reflections on Hydrology

Science and Practice

Nathan Buras
Editor

American Geophysical Union, 1997

Published under the aegis of the AGU Book Board
Richard Vogel, Hydrology Editor

Library of Congress Cataloging-in-Publication Data

Reflections on hydrology : science and practice / Nathan Buras, editor.
p. Cm.
Includes bibliographical references.
ISBN 0-87590-874-8
1. Hydrology. I. Buras, Nathan.
GB662.4.R43 1997
551.48--dc21 --dc21 97-43345
[551.48] CIP

ISBN 0-87590-874-8

Printed in the United States of America

2000 Florida Avenue, N.W.
Washington, D.C. 20009

Contents

Courtesy of University of Arizona

Chester C. Kisiel
1929-1973

At the time of his death, Chester Kisiel was a professor at The University
Arizona, holding appointments in both the Department of Hydrology and Wat
Resources and the Department of Systems and Industrial Engineering. Kisi
came to Arizona in 1966 from the University of Pittsburgh, where he taught
the Civil Engineering Department from 1954 to 1965. The theme of h
professional work was his continuing effort to bring mathematical and mode
engineering methods to bear on problems in hydrology. The fruits of his effor
are evidenced in his many publications, in several international symposia
which he played a leading organizational and scientific role, and perhaps mo
important of all, in the stimulation and guidance he gave to his colleagues a
students. By these efforts he established an international reputation and a de
personal esteem on the part of those with whom he collaborated. The series
lectures and, by extension, this volume are dedicated to his memor

BIOGRAPHY

CHESTER C. KISIEL

JULY 9, 1929 - NOVEMBER 5, 1973

Chester Kisiel was born in Harrison township, the eldest of six children. He came from an immigrant background, from the steel mills of Pittsburgh. Chester had to work from the time he was fourteen years old to help support himself and to help pay his educational expenses, and he never stopped working until his untimely death on a handball court in Tucson.

At the time of his death, Chester Kisiel was a Professor at The University of Arizona, holding appointments in both the Department of Hydrology and Water Resources and the Department of Systems and Industrial Engineering. He came to Arizona in 1966 from the University of Pittsburgh, where he taught in the Civil Engineering Department from 1954 to 1965. Professor Kisiel was educated in Civil and in Sanitary Engineering at Pennsylvania State University and at the University of Pittsburgh. He received the degree of Doctor of Science from the latter institution in 1963. He was in the U.S. Air Force, 1951-53, serving in Japan and Korea.

There was a theme to Professor Kisiel's professional work — it was his continuing effort to bring mathematical and modern engineering methods to bear on problems in hydrology. And he was not content to deal with existing problem statements. In many cases he refined and reformulated the problem itself, or he identified problems before many of his colleagues were aware of their existence. The fruits of his efforts are evidenced in his many publications, in several international symposia in which he played a leading organizational and scientific role, and perhaps most important of all, in the stimulation and guidance he gave to his colleagues and students. By these efforts he established an international reputation and a deep personal esteem on the part of those with whom he collaborated.

Chester Kisiel had the gift of self-examination, which is another way of saying that he had the gift of honesty. He tried to be honest with himself and honest with others. He could forgive many things but not something, that in his view, was a dishonest piece of work.

Professor Kisiel brought to bear prodigious gifts in pursuing his goals. He had the gift of hard work and uncompromising standards. He was a hard task master, but he never demanded more of others than he was willing and able to do himself. He had the gift of sound instinct, both with regard to technical matters and in the assessment of the strengths of his colleagues. He had the gift of stimulating and working with others across many disciplines.

Preface

In the early 1980s, the Department of Hydrology and Water Resources at th University of Arizona started a tradition: an annual public lecture to perpetuate th memory of one of its most original thinkers who passed away at an early age, Chest C. Kisiel. At that time, the department was quite young — a little over ten years old – and so was the University of Arizona, not quite a century old. The overall atmosphe was extremely stimulating, faculty members and students were curious and excite wishing to learn and understand more about the natural phenomena that transfor precipitation into water and the possible development of regional waters for huma uses. The preparation and delivery of these lectures were entrusted by the departme to outstanding scientists in the fields of hydrology and water resources, thus attainin a double objective. On the one hand, the lectures became salient points on a tim trajectory when specific facets of the broad agenda of scientific issues studied in th department were brought to the limelight of a public discourse. On the other hand, th lectures also provided opportunities for reflection on contemporary problems and on th approaches for their study and analysis.

The study of natural phenomena, such as runoff, streamflow, infiltratio evapotranspiration, can be done in a strictly descriptive, non-quantitative way. Thi however, is hardly satisfactory because it does not offer a basis for quantitativ decisions. For this reason, we prefer to describe hydrologic phenomena usin mathematical models representing functional relationships between the interactiv factors. Thus, mathematical models were developed especially for various parts c hydrology, in addition to borrowing analytical models and approaches from othe scientific disciplines and adapting them to problems in hydrology and water resource All the distinguished lecturers used, presented, discussed and explained models of broad range of complexity.

The fundamental problem in water resources is water management, i.e., the decision we make regarding the storage of water, its conveyance to centers of use, its treatme and the handling of the discharged waste water. This activity requires primarily th collection of information — hydrological and non-hydrological — and its processin so that alternatives can be identified and ranked according to preferences, enablin decision makers to select the preferred alternative. Two closely related problems ar paramount in managing water resources: (1) how much and of what quality of th resource is left for future generations; (2) what is the response of the environmen primarily in terms of water quantity and its quality, to management stresses? Thes issues have appeared during the last three decades on the horizon of the hydrologica sciences.

To start unravelling these problems, we need to know and understand more about th interactions between atmosphere, lithosphere, hydrosphere and biosphere on a variet of scales. The attainment of this knowledge and understanding will require close ooperation of surface hydrologists, hydrogeologists, meteorologists and environmental cientists. Of course, global scale hydrology will give us only a rough idea of possible egional changes which may be caused by global climatic fluctuations. One must ealize, however, that semi-arid and arid regions are considerably more sensitive to iese fluctuations than semi-humid and temperate regions.

The matter of scale, therefore, becomes important, both regarding time and space. ombining length, one of the dimensions of space, and time in an appropriate fashion, e obtain velocity. Indeed, velocity is an essential factor in the hydrological cycle hich describes the continuous motion (perpetuum mobile?) of water on planet Earth. his motion, however, occurs at different velocities in different segments of the cycle. he best illustration would be the velocity of streamflows on the surface of the earth ompared with the movement of water in aquifers below the land surface. Two bservations come to mind immediately: one is that different hydrologic laws may be iscovered at different time and space scales, the other is that we have been unable so ar to produce one completely satisfactory mathematical model of a regional water ystem consisting of both surface streams and aquifers, and it is doubtful that we will ver will, due to the differences in space-time scales which control these velocities.

During the quarter-century since the passing away of Chester Kisiel, the emphasis in ie broad field of "water resources" shifted from problems centered on water quantity nd availability to the quality of water and its management. Since water bodies were sed for a long time as conveyors and/or repositories of waste, primarily because of the ieir natural capability of "disposing" of pollutants, water quality problems became nportant quite early in history. Their resolution was entrusted very early to the hands f the same professionals who supplied so efficiently the quantities of water the opulation wanted: the civil engineers. Engineering solutions were effective to a certain oint, beyond which they essentially shifted the problem of degraded water quality from here it was offensive to somewhere else, where another group or individual had to andle it. One of the main reasons of this non-solution was the rarity of environmental iicrobiologists involved in the study and resolution of water quality problems. ortunately, this situation is slowly being redressed, and bioremediation of ontaminated soil and water is becoming an accepted procedure for water quality ianagement.

The pioneering scientists and engineers who established one of the fundamental onceptual structures in the early 1960s were proud, and rightly so, that their efforts erformed the "marriage of engineering, hydrology and economics." One of the nportant concepts adopted then was that of economic efficiency, meaning that a water esources development project when implemented will create a situation where no one n the area served by the project will be worse off than before its implementation. ndeed, this seems to be a worthwhile objective of water resources development rojects. However, one should also consider that every economic decision has a social imension. Attainment of economic efficiency, in most cases, does not affect positively ssues of equity; quite contrary, the regional distribution of income may be worsened.

Furthermore, socio-economic and political issues transcend economic efficiency and equity and include also the degree of participation (if any) of the users of water and of the customers of a regional water resources development project in the planning and management of the project. The study of this dimension has started only recently.

As we use increasingly sharper analytical methods in the study of hydrological and water resources issues, the matter of water storage over specified periods of time acquires increasing importance. One critical factor in this matter is the vadose zone of the soil. This unsaturated stratum has two important functions: it is capable of storing significant amounts of water between irrigations; and it supplies the growing crops with the moisture necessary for their yields. Recognizing these capabilities of the vadose zone, one can begin planning irrigation projects and their management not for maximum yield within a short horizon but for sustained production over a long period of time. Expressed in broader terms, we can say that the demand for water in the agricultural sector is not a sanctified amount but a quantity that varies with the type of crop grown, its physiological level of development, climatic fluctuations, and the properties of the vadose zone of the soil where the crop is grown.

Because of the continuous increase of the world population while the planet's resources are fixed, a situation of water scarcity is developing in several regions. The scarcity may be due not only to increased demands for the amount of water available in a region, but also because the quality of some of the regional water resources is degraded. The quality of regional waters is also a matter of dynamic equilibrium. The degradation of water quality is due to the discharge of material and energy wastes into water bodies, even where no appreciable anthropogenic activity occurs. At the same time, populations of microorganisms composed of many genera and species use the water pollutants as sources of nutrients and energy for their own metabolical processes, thus decomposing them into non-toxic compounds. Environmental microbiology is thus an important scientific discipline for the management of water quality.

Coming back to the phenomena in nature where the hydrosphere, atmosphere, lithosphere and biosphere interact, we observe that the phenomena are generated by the continuous stream of energy flowing from the sun. This energy produces fluctuations in the environmental temperature, the precipitation of atmospheric water, and the movements of the atmosphere — the winds — in one word, the climate. We may consider then the climate to be the forcing function of the hydrological phenomena and the lithosphere, through its geological formations, to be the basic constraining factor of these phenomena. Overlapping these natural factors are the anthropological interventions, so that in many parts of the world man — generically — is a major modifying agent.

Perhaps the most significant human intervention in the hydrological cycle is that which resulted in the degradation of the quality of the aquatic environment. Whenever and wherever humanity appeared on the globe, it joined the already existing flora and fauna in discharging their metabolic wastes into the environment. An equilibrium was probably attained with the existing microbiological populations so that these pollutants were recycled as non-harmful compounds. The rapid industrialization of the world in the last quarter-millenium produced contaminants that did not exist before, many of iem synthesized for specific industrial processes, and that may be highly toxic to umans (some could be carcinogenic). We face again the strong necessity of increasing ur knowledge and understanding of environmental microbiology.

The last lecture included in this volume returns almost full circle to the central issue round which "water resources" revolves: storage of water and the management of orage facilities. Historically, water storage was considered to occur on the surface, in eservoirs create by dams. Sometime in this century, we began considering aquifers as orage components of regional water resources systems. Recently, we started to realize iat the economic sector using about 70% of the water developed in the world — griculture — has an additional storage component, the root zone of the soil. Designing ater storage facilities and managing them, particularly on the surface, are expensive id complex projects. The complexity stems primarily from the fact that when perating a storage facility two related decisions have to be taken simultaneously at oecific points in time: how much water to release from storage, and how much water leave in the reservoir. The mathematical analysis of the problem did not yield a set f independent equations which would give a unique solution. As a result, water storage ay be either a blessed inspiration contributing to socio-economic development, or a ource of continuous desperation generated by insufficient information and erroneous ialysis.

Nathan Buras
May 1997

FOREWORD

It is hoped that the first Chester C. Kisiel Memorial Lecture starts a tradition. The Department of Hydrology and Water Resources is a young department, part of the University of Arizona—also a relatively young university; thus it is particularly gratifying to plan and experience an event which is meant to recur periodically and so to become an integral part of our academic life.

There is a double objective in establishing the Chester C. Kisiel Memorial Lectures. On the one hand, they should become salient points along a trajectory in the time dimension when specific facets of the broad panoply of scientific issues studied by this Department will be brought into the limelight of public discourse. On the other hand, the Memorial Lectures will provide opportunities for taking pause and reflect upon problems faced at the time and upon approaches and methodologies used for their study and analysis. Both these objectives were fully attained by the first Lecture.

The natural and social phenomena which are the domain of Hydrology and Water Resources are very complex and necessitate a variety of approaches for their understanding. These approaches range from strictly deterministic cause-effect relationships to totally random considerations. Within these two extremes lies a very difficult intermediate terrain. A quotation from Max Born will serve to illustrate the philosophical problem:

> *"An unrestricted belief in causality leads necessarily to the idea that the world is an automaton of which we ourselves are only little cog-wheels. This means materialistic determinism...On the other hand, an unrestricted belief in chance is impossible, as it cannot be denied that there are a great many regularities in the world; hence there can be, at most, a 'regulated accident'. One has to postulate laws of chance which assume the laws of nature or laws for human behavior...Our philosophy is dualistic in this respect; nature is ruled by laws of cause and laws of chance in a certain mixture."*
>
> Max Born, Waynflete Lectures

The first Chester C. Kisiel Memorial Lecture addresses this issue squarely.

Nathan Buras

REFLECTIONS ON HYDROLOGY

N. C. Matalas
U.S. Geological Survey, Reston, VA 22092

It was, as I recall, in the early spring of 1965 that Professor John Harshbarger and I were taking a flight from Tucson to Phoenix for quite unrelated reasons. Shortly after take-off, one of the two engines of the plane stalled and that event prompted our immediate return to Tucson. The wait for another flight gave us the opportunity, that otherwise we would not have had, to converse. Our attention soon turned to the University of Arizona's Hydrology Program that Professor Harshbarger chaired—a program that had begun in 1961 with the support of the U.S. Geological Survey. Professor Harshbarger inquired if I knew of anyone who might be interested in a position within the Program. One whom I mentioned was Chester Kisiel who was, at the time, an assistant professor of civil engineering at the University of Pittsburgh. Subsequently, Chester received and accepted an offer from the University of Arizona and served with distinction on its faculty until his untimely death in 1974.

If my recollection serves to explain, in part at least, how it was that Chester came to Arizona, then our attention should focus not on Professor Harshbarger's question nor on my response, but on the event that led to our conversation—the stalling of one of our plane's two engines. If we regard that event, a single case, as having been a chance outcome of a set of circumstances, then we might, upon further elaboration, interpret the explanation of how Chester came to Arizona as being a statistical one.[1] The substance, more so than the specifics, of that sort of explanation would have been of interest to Chester; it is this kind of explanatory consideration which so often arises in our hydrologic studies. That we do not fully appreciate the scientific import of this kind of situation—namely, a single case with a probabilistic dimension—is primarily attributable to the operational emphasis that is currently placed on hydrologic research, an emphasis which encourages a deterministic notion of causality. This is somewhat of a paradox, as the single case is so characteristic of operational problems that entail the making of decisions in a state of uncertainty. Thus, in keeping with Chester's interest, let us reflect on the notion of chance within the context of hydrologic research.

In the United States, emphasis in hydrologic research is placed on algorithmic solutions to operational problems of interest to management. Our more current literature attests to this emphasis; for the most part, the papers address problems of estimation, broadly construed, and of optimization. While in the United States public attention for hydrology is drawn to engineering accomplishments,[2] on the European continent care was taken to distinguish between scientific and applied hydrology.[3] Thus in contrast to the American literature, the European literature suggests a greater commitment to hydrologic description and explanation that is in keeping with a natural science tradition. And in particular, the Russians have expressed far greater interest in defining global water balances by which a fuller account may be provided to the dynamics of the hydrologic cycle.[4] It is noted that by hydrologic description is meant a description of the dynamics of the system and not just a simple statistical summarization of the data; the latter is both of limited scientific content as well as of limited management utility.

It is interesting to note, that efforts in the United States directed towards establishing a more direct relation between water management and land management effectively call for a more geomorphologic perspective of hydrology. Moreover, efforts to accommodate environmental and water quality concerns within the general scope of water quantity management call, in effect, for meaningful water balances on a regional, if not on a global, scale.

The problems of estimation and optimization derive from operational interests in predicting future hydrologic events of a kind which are well-known, e.g., runoff, flow and transport, drawdown, oxygen depletion and re-aeration and water pricing. Prediction of this kind presupposes a theory whose operational context is known in advance: the rejection of the theory's predictions does not imply its falsification but merely serves to define its limit of applicability, e.g., when the cost of reducing prediction errors exceeds the benefits derived from the predictions. Such predictions are to be distinguished from those which are of scientific interest—those which are deduced from the implications of a theory. Such predictions and thereby the theory from which they are deduced can only be indirectly tested by a critical experiment, with which, if it constitutes a counterexample, the theory may be falsified.

Basic to the making of predictions of either kind is the systematically collected portion of the extant hydrologic data base which has evolved in major part in response to the questions of management rather than having been motivated by scientific inquiry, a situation not unique to the United States. If one adopts a strictly positivistic viewpoint, then the principal aim of scientific endeavor is to develop and enhance predictive capability conditioned on observations of physical and social phenomena. Thus, one might adopt the view that the purpose of data collection programs is to develop the set of observations, i.e., data summaries, to which scientific endeavor may be applied to produce better predictive capability of the occurrence of such hydrologic phenomena as floods, droughts and eutrophication.

However, any specific configuration of stations that is imposed for the collection of hydrologic data defines the set of observations that will be obtained in terms of externally imposed conditions, e.g., most "efficient" in producing "information" when the measures of efficiency and information are defined within the operational scope of network design, rather than internally imposed conditions, such as those dictated by the data as in "event" sampling. Thus, if predictive capability arises from scientific endeavor conditioned on the observations and if the limitations to the set of observations are externally imposed, then the degree to which predictive capability may be developed and enhanced is also limited by these external constraints, rather than by constraints imposed by the scientific endeavor itself. That is, the design of the network has served as a constraint on the science although complying with the operational request.

Furthermore, even if one adopts a broader view of scientific endeavor than positivism allows, namely that theory precedes observations and dictates what observations are to be made to evaluate the consequences of the theory and thereby to evaluate the theory itself, it should be recognized that there is no way of empirically evaluating the results of network design, namely the configuration of stations and sampling schedule imposed. To the extent that designing the network is a scientific endeavor, network design is a logical construct as its results can be evaluated in terms of their logical consistency but they cannot be empirically tested.

Unless one recognizes that network design is a logical construct rather than an empirical science, one will seek an empirical rather than a logical assessment as to whether or not the particular design of a network meets the objectives that were intended to be met. But even if network design is recognized as a logical construct, it must also be recognized that the objectives to be met by network design[5] derive from questions that to some degree are transscientific[6] and cannot therefore be satisfactorily addressed by this endeavor alone.

The problems of estimation and optimization derive from questions of how best to manage the Nation's water resources. Answers to those questions are very much in the public interest and there is no doubt that those questions merit our attention. But must they exclude any research directed towards other hydrologic questions? Obviously not. However, the scope of the current hydrologic literature suggests that those problems are claiming the major share of our research attention. We have effectively limited our scientific interests by placing matters of operational expediency—providing algorithmic and computational techniques convenient for addressing water management issues—ahead of concerns over hydrologic reality—questioning the empirical consistency of the assumptions that have been imposed in order to develop those operational techniques. Some examples of controversial scientific issues on which the discussion has been curtailed by operational rather than scientific arguments serve to illustrate this compromise.

The perception of hydrology as contextually residing within the low frequency domain of the theory of turbulence[7] has not been pursued following the earlier promise that the Hurst phenomenon[8] might be explained in terms of stationary stochastic processes of infinite memory. The lack of interest in exploring such an original proposal may in large part be attributed to the argument that management decisions are relatively insensitive to the extent of hydrologic memory, and that therefore the algorithmic and computational convenience afforded by short memory stationary stochastic processes in addressing operational problems more than compensates for whatever degree of hydrologic reality such processes fail to express. The basis for accepting such an argument is a viewpoint advanced through the general scope of benefit-cost analyses relating to the management of water resource systems, namely that the uncertainties attendant to economic concerns have a greater bearing on management decisions than do uncertainties regarding the characterization of physical processes.

However, the strength, if not the validity, of this viewpoint may be questioned on the probabilistic grounds upon which it is predicated. Economic uncertainty in water management is generally addressed by statistical decision theory[9] predicated on a personalistic interpretation of probability, a subjective interpretation of probability[10] whereas uncertainty in regard to physical processes is measured in terms of relative frequency[11] which is, in effect, an estimate of limiting relative frequency,[12] an objective interpretation of probability. These are distinctly different interpretations of probability—the former being a measure of our degree of belief about the hydrologic system and the latter being a limiting characteristic of an infinite sequence of events of a recurrent kind. Thus little meaning can be given to the ratio of subjective to objective measures of probability as it is a comparison of noncommensurables. If the ratio is to have some meaning, we must either measure the uncertainty about physical processes by subjective probabilities or measure the uncertainty over economic matters by objective probabilities. To choose to do the former is to yield on objectivity in the course of our hydrologic research.

Even so, the viewpoint that economic concerns dominate those associated with physical processes is of little significance in the face of substantive claims that the water resources of the Nation are rapidly approaching a state of high development. As that state is approached, the range of the cumulative departures from the mean becomes a more meaningful design variable,[13] such that the uncertainty regarding the characterization of physical processes becomes more nearly synonymous with the economic uncertainty itself. Thus, if there is an argument, it is over the interpretation of the measure of uncertainty, i.e., of probability. And that argument would be over how we objectively interpret probability if we would direct our scientific as well as operational attention to learning more about the hydrologic system than about our degree of belief about the behavior of the system. And with that direction of our scientific and operational interests, we should be motivated to question the assumption which we have imposed in developing methods to provide for algorithmic and computational convenience in addressing water management issues.

A second example of the compromise of research to operational interests can be found in estimating flood quantiles, where we have restricted our choices of probability distributions to those which are unbounded above. Thereby we have assumed, in effect, that there is no limit on the magnitude of a flood. Though this assumption provides for algorithmic convenience in the statistical estimation of flood quantiles, it has little in the way of hydrologic support; indeed it can be argued that the carrying capacity of a drainage basin is infinite.[14] It is often said that the assumption of an unbounded probability distribution of floods is good enough for all practical purposes. But we really don't know that it is so and that itself is reason enough to question the assumption on hydrologic grounds.

Yet another example arises in connection with the classical theory of extreme values.[15] The earlier prospect that this theory might provide theoretic support to the estimation of flood quantiles has not been pursued with much enthusiasm. The decline in hydrologic interest in the theory may be attributed to the narrowness with which distributions arising naturally from the theory have been applied. For a time, one of the asymptotic distributions attaining under the theory, namely that referred to as Type I or Gumbel, was used fairly extensively in flood studies explicitly as the distribution to describe flood quantiles. That particular distribution is unbounded above, as well as below, and has a fixed shape. However, any number of multi-parameterized distributions whose shapes are variable can be found to provide a closer fit to flood data. This has prompted not only the choice of distribution to be restricted to those which have variable shape as well as being unbounded above, but also has led to a loss of hydrologic interest in the theory of extreme values.

It is interesting to note that under the classical theory of extreme values with the assumption of stochastic independence, the Type I distribution may attain even if the initial distribution is bounded above.[16] This does not suggest that the Type I distribution is a rational hydrologic choice for estimating flood quantiles, but it does suggest that in the course of questioning the assumption of no limit on the magnitude of a flood we do re-examine extreme value theory, particularly its more recent extension in which the assumption of stochastic independence has been relaxed.[17] If this re-examination is to lead to theoretic support for the statistical estimation of flood quantiles, an acceptable empirical, i.e., hydrologic, interpretation will have to be made of the mathematical terms used in the theory of extreme values. This further suggests that whatever hydrologic interest we may have in a particular mathematical theory or construct is not likely to be sustained if we do not or if we cannot empirically interpret the intrinsic terms of that theory or construct. It is perhaps because of the lack of such interpretation that hydrologic interest in Kalman filtering has quickly waned.[18,19]

Finally, as the Nation's water resources are stressed, water management issues are thought to have become more complex: the dimensionality of the issue is perceived to be increasing as more variables are to be considered. For lack of structured, empirically testable, hydrologic theories which allow us to deal with what is viewed as not only an open, but an expanding system, we have turned, with marked success, to an approach which allows the incorporation of finite, but unspecified, large numbers of variables in the formulation of operational problems, namely, systems analysis.[20] In order to take this approach, we must make strong hydrologic assumptions—many of which are never re-examined after they have served to "solve" the problem—but by making these limiting assumptions, we can link together as many variables as the operational problem seemingly requires as long as the variables can be expressed in commensurable terms. However, by the very explicit method of treating linkages—which is the basis of the success of these methods—the broad adoption of a systems-analytic approach to addressing water resources questions has encouraged the fragmentation rather than a synthesis of hydrologic knowledge.

There are two other elusive ramifications that have followed from the success of the application of these techniques. By allowing the addition of new variables with each new statement of a water resource problem, systems analysis has encouraged the belief that there will always be more hidden variables with which we need not cope simply by the application of our primitive hydrologic theories. And although management may accept the results of such analyses, we can only confirm, at best, the logical consistency of the problem formulation, not its empirical consistency. That is, we have removed the requirement of empirical accountability from our scientific work.

Such examples as the above are more the rule than the exception. It can be said that to a large extent hydrologic research is primarily externally directed by operational concerns rather than internally motivated through scientific pursuits themselves. As the Nation's water resources are further stressed, the research community will be further pressed to respond to external directions of immediate operational problems that will occur both inevitably and more demandingly with time. But if the operationalism of hydrology continues to be further emphasized, research programs risk becoming absorbed by or essentially mere extensions of operational programs.

I am not implying that hydrologic research should or must be conducted strictly outside the context of operational problems that are of immediate concern to water management. Scientific issues can be traced to problems of a very practical kind, though not easily in reference to the mature sciences. But hydrology is not a mature science and so we should not expect our present scientific concerns to be much different in substance and form from the more immediate problems confronting operational hydrology. Our expectation is that in the not too distant future, hydrologists will have to look deeper to find the practical roots of their scientific concerns. The realization of that expectation is as much in the public interest as in the self interest of hydrologists, for the obvious reason that there be a firmer scientific foundation upon which to resolve water management issues.

The internal direction of hydrologic science and hence, of research, is towards the goals of explanation and description of the connectedness between events in terms of axiomatic theories within the conception of the hydrologic cycle, as well as towards discerning the consequences of those explanations and descriptions. The difference between the internal direction of research and the external direction attributed to it by operational programs gives rise to a mutual non-complementarity between research and operational endeavors. But only by the internal direction can scientific knowledge grow and, paradoxically, only by this growth can research support operational programs. Thus to repeat, continued growth of our scientific hydrologic knowledge is no less in the public interest than is the resolution of immediate operational problems in regard to the management of the Nation's water resources.

If we are to further the realization of our expectation of a mature science of hydrology, we will have to give somewhat less than our undivided attention to the resolution of issues that are of immediate concern to water management. And this we must do, if for no other reason, to better support operational programs. Some risk is entailed in following this course; we cannot guarantee the success of specific scientific endeavors. The tasks of reducing standard errors and lessening sub-optimizations are in one way or another of real or perceived importance to the management of water resource systems. Though the tasks are demanding, they are in a sense scientifically riskless; we can always improve on our past performance and claim that improvement reflects the relevance of research to operations. The fact that management so readily accepts that claim has embued us with a large degree of scientific conservatism. However, if we are to further the realization of our expectation of a mature science of hydrology, we will need to take a risky course—a scientific course. With realism and objectivity as our guide, the risk would not be so large.

I cannot offer a definitive prescription for assuring the continued growth of our hydrologic knowledge with due attention to resolving immediate operational problems in regard to the management of water resource systems. But as I asked that we give somewhat less than our undivided attention to the resolution of issues that are of immediate concern to water management, I feel obligated to offer some suggestion of what we might begin to do with our freed time. So let me suggest for our consideration a newer objective interpretation of probability—propensity—and close with some brief remarks on what might be hydrologically gained with that interpretation of probability.

A prevailing viewpoint in hydrology is that "the occurrence of all natural phenomena will be found based on law and order, if one can only analyze the conditions surrounding these phenomena and evaluate the varying influences and effects of these conditions in each instance."[21] This is the problem of the single case which so often arises in both questions of scientific and operation interest—the problem of explaining the occurrence of a single event or a realization of a process, e.g., the contamination of a local water supply, the eutrophication of a lake or the change in the course of a river. Such problems have frequently been described by the statement that "every river is, in effect, a law unto itself."[22] For the most part these problems have been perceived as devoid of probabilistic content and are addressed with a deterministic notion of causality, a traditional approach[23] which is now referred to as physics-based modeling when dealing with those problems having strictly a water quantity dimension.

Nonetheless, many of the problems can in a more-or-less meaningful way be perceived as the single outcome of an experiment, controlled or not, and that suggests, if only intuitively, that some account of probability should be taken in their explanation.[24] With this viewpoint, an approach based on the personalistic interpretation of probability, generally referred to as Bayesian analysis, has been recently introduced in hydrology.[25] This approach is in outward appearance different from the traditional approach, yet both approaches are guided by a deterministic notion of causality and they are complementary—the limitation of the traditional approach due to the lack of knowledge or information is expressed by this newer approach as a degree of belief or a degree of knowledge.[26] This personalistic approach can take us no further than the traditional approach and there is reason to believe that that may not be very far.

Our experience suggests that hydrological events are of a recurrent kind and are the realizations of hydrologic processes. And our intuition tells us that the events are chance outcomes of experiments, the collection of processes themselves, and that chance has something to do with the experimental arrangements, particularly with their initial conditions. However, knowing the initial conditions is not tantamount to knowing what the outcome of the experiment is to be. Thus, intuitively, probability is an intrinsic empirical property of the system. Our intuition would not serve us too well if the outcome of the experiment, a single case, is uniquely accountable with a subjective interpretation of probability, if it is to be probabilistically accountable at all. However, the single case can be addressed by objective probability, perhaps not by the limiting relative frequency interpretation of probability but by a closely related interpretation, that of propensity—the disposition or tendency for a system to yield an outcome of probability p given the arrangement,[27] i.e., the chance set-up, of the experiment, i.e., the system. If nothing else, the notion of chance set-up serves to remind us that probability statements are statements of conditional probabilities. Thus, just as the probability of a flow event is conditioned on the physiographic and meteorological arrangements, a probabilistic description of a water quality event must be conditioned on the natural and manmade arrangements under which the event has or may occur.

Propensity probabilities can interact and combine, just as forces can interact and combine in physical systems. The analogy between the propensity probabilities and forces operating on a physical system suggests that the conventional physical interpretation of the hydrologic cycle[28] should rather be modified to one of a hydrologic system, more in keeping with the notion of chance set-up. If the hydrologic system is the chance arrangement of physical processes, then a broader conception of the hydrologic cycle is needed, one which recognizes human activity as inseparable from the natural system.[29] The structural as well as nonstructural measures of water management join with the natural conditions to constitute the chance set-up from which the propensities of the hydrologic system arise; hence, necessarily the hydrologic system must be conceived to incorporate human activity.

With this conception of the hydrologic system as the chance set-up, we need measures to allow discussion in commensurable terms of the diverse variables of a water resource system, including physical, chemical, biological as well as economic considerations. One such measure may be provided by the concept of entropy which arises in the study of the thermodynamics of systems, is familiar to biology through measures of species diversity[30] and has been introduced to economics through the idea of the dispersion of natural resources.[31] One general problem which arises from the adoption of this expanded notion of the hydrologic system is to account for the propensity of the system at a given time to have entropy e with probability p, which becomes a way of addressing the orderliness of a closed system. The inadvertent actions of man and the vagaries of nature lead to an open system. With this broadened conception of the hydrologic system, it is the aim of management, through the expenditure of energy, to alter the chance set-up such that the hydrologic system is closed and has the propensity for a reduction ê e in entropy with probability p.

And now to close. Of what I have said, that which is creditable, I must share with Jurate Landwehr; the rest is mine to bear. I don't know just what Chester would have found creditable. But I have no doubt that he would have posed substantive questions. And what is there to be gained, if we never question our hydrologic paradigms?

SELECTED BIBLIOGRAPHY

1. Salmon, Wesley C., with contributions by Richard C. Jeffrey and James C. Greeno, *Statistical Explanation and Statistical Relevance*, 117p, University of Pittsburgh Press, Pittsburgh, PA, 1971.
2. Meyer, Adolph F., *Elements of Hydrology*, 487p, John Wiley and Sons, Inc., New York, NY, 1917.
3. Meinzer, Oscar E., *Physics of the Earth-IX: Hydrology*, 712p, McGraw-Hill Book Company, Inc., New York, NY, 1942.
4. L'vovich, M. I., *World Water Resources and their Future*, Mysl P. H., Moscow, USSR, 1974, Translation by the American Geophysical Union, edited by R. L. Nace, 415p, Washington, D.C., 1979.
5. AGU Chapman Conference, Design of Hydrologic Data Networks, Tucson, Arizona, December 11-14, 1978, Proceedings in Water Resources Research, V. 15, No. 6, pp. 1673-1871, December, 1979.
6. Weinberg, Alvin M., Science and Transscience, Minerva V. X, No. 2, pp. 209-222, April 1972.
7. Mandelbrot, B. B. and Wallis, J. R., Noah, Joseph, and Operational Hydrology, Water Resources Research, V. 4, No. 5, pp. 909-918, October, 1968.
8. Hurst, H. E., Long Term Storage Capacity of Reservoirs, Trans. Am. Soc. Civil Engr., V. 116, Paper No. 2447, pp. 770-799, 1951.
9. Raiffa, Howard and Robert Schlaifer, *Applied Decision Theory*, 356p, Harvard University Press, Cambridge, MA, 1961.
10. Savage, L. J., The Foundations of Statistics, 310p, Dover Publishing Company, New York, NY, 1972.
11. Hazen, Allen, Storage to be Provided in Impounding Reservoirs for Municipal Water Supply, Trans. Am. Soc. Civil Engr., V. 77, Paper No. 1308, pp. 1539-1640, 1914.
12. von Mises, Richard, *Probability, Statistics and Truth*, 241p, George Allen and Unwin Ltd., London, England, 1957.
13. Wallis, J. R., and Matalas, N. C., Sensitivity of Reservoir Design to the Generating Mechanisms of Inflows, Water Resources Research, V. 8, No. 3, pp. 634-641, June, 1972.
14. Horton, Robert E., Discussion of Flood Flow Characteristics by C. S. Jarvis, Trans. Am. Soc. Civil Engr., V. 88, Paper No. 1589, pp. 1081-1086, 1925.
15. Gumbel, E. J., *Statistics of Extremes*, 875p, Columbia University Press, New York, NY, 1958.

16. Gnedenko, B. V. Sur la Distribution Limite de Terme Maximum d'une Série Aléatoire, Annals of Mathematics, V. 44, pp. 423-453, 1943.
17. Galambos, Janos, *The Asymptotic Theory of Extreme Order Statistics*, 348p, John Wiley and Sons, New York, NY, 1978.
18. AGU Chapman Conference, Applications of Kalman Filter to Hydrology, Hydraulics and Water Resources, Pittsburgh, Pennsylvania, Proceedings, 783p, Published by Stochastic Hydraulics Program, Department of Civil Engineering, University of Pittsburgh, Pittsburgh, PA, 1978.
19. Schrader, B. P. and S. F. Moore, Kalman Filtering in Water Quality Modeling: Theory and Practice, Winter Simulation Conference, National Bureau of Standards, Rockville, MD, December 5-7, 1977.
20. Maass, Arthur et al, *Design of Water-Resource Systems*, 620p, Harvard University Press, Cambridge, MA, 1962.
21. Meyer, Adolph E., Op. Cit. p. 2.
22. Rafter, George W., Relation of Rainfall to Run-off, Water Supply and Irrigation Paper No. 80, U.S. Geological Survey, Washington, DC, 1903.
23. Mead, Daniel W., *Notes on Hydrology*, 202p, Shea Smith and Co. Press, Chicago, IL, 1904.
24. Rafter, George W., Hudson River Storage—Second Report on Survey of Upper Hudson Valley, Annual Report of the State Engineer and Surveyor, Appendix 8, pp. 803-858, Albany, NY, 1895.
25. International Symposium on Uncertainty in Hydrologic and Water Resources Systems, Sponsored by the National Science Foundation, Washington, D.C., Tucson, AZ, December 11-14, 1972.
26. Giere, Ronald, Objective Single-Case Probabilities and the Foundations of Statistics, in *Logic, Methodology and Philosophy of Science*, IV, edited by P. Suppes et al, pp. 467-483, North-Holland Publishing Company, Amsterdam, The Netherlands, 1973.
27. Popper, Karl R., The Propensity Interpretation of Probability, British Journal of the Philosophy of Science, V. 10, pp. 25-42, 1956.
28. Horton, Robert E., The Field, Scope, and Status of the Science of Hydrology, Trans. American Geophysical Union, pp. 190-202, 1931.
29. Matalas, Nicholas C., Jurate Maciunas Landwehr and M. Gordon Wolman, Prediction in Water Management in *Scientific Basis of Water Resources Management*, pp. 118-127, National Research Council Geophysics Study Committee, National Academy of Sciences, Washington, DC, 1982.

30. Pielou, E.C., The Use of Information Theory in the Study of the Diversity of Biological Populations, *Proceedings of the Fifth Berkeley Symposium on Mathematical Statistics and Probability*, pp. 163-177, University of California Press, Berkeley, CA, 1967.
31. Georgescu-Roegen, Nicholas, *The Entropy Law and the Economic Process*, 457p, Harvard University Press, Cambridge, MA, 1971.

NICHOLAS C. MATALAS

U.S. Geological Survey, Reston, VA 22092

Dr. Nicholas C. Matalas has been a research hydrologist with the Water Resources Division of the U.S. Geological Survey for more than 25 years. In recognition of his research in network design, hydrologic time series, flood frequencies and regionalization of hydrologic information, he has received the Meritorious Service Award and the Distinguished Service Award from the Department of the Interior and the Horton Award from the American Geophysical Union. He served as President of the Section of Hydrology, American Geophysical Union and is a Fellow of the American Geophysical Union.

Over a period of seven years, Dr. Matalas served as Chief of the Water Resources Division's Systems Analysis Group. During this time he directed studies in assessing the management of the Outer Continental Shelf mineral leases by the U.S. Geological Survey, the effectiveness of the Bureau of Mines' implementation of the 1968 Coal Mine Health and Safety Act and residuals management and water use in coal gasification processes.

He received his BS degree in Civil Engineering and MS degree in Sanitary Engineering from North Carolina State College. And under Professor Harold A. Thomas, Jr., he received his PhD degree from Harvard University.

As U.S. Geological Survey support in beginning the Hydrology Program at the University of Arizona, Dr. Matalas, along with John Ferris and Herb Skibitzke, served on the Program's faculty during the seminal academic year, 1961-62.

FOREWORD

The second half of the decade of the fifties saw a burst of activity in the field of hydrology and water resources. Independent efforts at different universities in the United States increased our understanding of hydrological processes and adapted mathematical methods developed in other areas of scientific endeavor to the analysis of water resource systems. The Second Kisiel Memorial Lecture captures, in a sense, the kernel of the quantum jump which occurred in the late fifties and early sixties in our perception of the complexities of the hydrological phenomena and man's relation to them. During that period, the late Chester Kisiel advanced our thinking in stochastic hydrology through his work at the University of Arizona; the Second Kisiel Memorial Lecturer was deeply involved in the Harvard Water Program; and the writer of this foreword made a modest contribution to the adaptation of some methods of mathematical programming to water resources systems analysis while at UCLA.

The time has come—a quarter century later—to pause and reflect on what has been done and, from this time perspective, ask ourselves how well did we do. In the sixties, young Ph.D.'s sallied forth from their universities into the real world spreading the good news that, at last, complex water resource problems in which imperfectly understood natural phenomena affected by anthropogenic interventions can be neatly disected by the application of systems analysis and the use of mathematical models. Some looked for problems that would fit their preconceived notions; others understood that the elegant conceptual structures spawned in an academic environment need additional forging in order to become tools that the professionals could use effectively in analyzing, understanding, and solving water-related problems. The Second Chester C. Kisiel Memorial Lecture directs a searchlight on this aspect of knowledge generation and its conversion into a practical instrument. But even more important than that, the Second Kisiel Lecture presents and discusses a philosophical basis which underlies the discipline called "water resources." The philosophical basis, together with the scientific foundation provided by hydrological sciences, are the conceptual framework within which the development and utilization of regional water resources take place.

Nathan Buras

THE REAL BENEFITS FROM SYNTHETIC FLOWS
Reflections on 25 Years With the Harvard Water Program

Myron B Fiering

The excitement of having been present at the Creation can last for a lifetime. You all recognize that science advances like a Bach fugue. First we have the innovation—the main theme—bold and brassy. This is followed by repetition of that theme in each of the several voices, corresponding to logical extensions of the motif within the scientific community. Then there is a period of free development with occasional reference to the fugal theme, corresponding to a long period of technological development and exploitation. And finally the main theme returns—a new idea, a new plateau is attained. It was my good luck to have been present at the Creation, a little more than 25 years ago, when the Harvard Water Program began systematically to consider ways of integrating engineering, economics and political science into the design of water-resource systems. I can use that last phrase with upper-case letters to identify the book and with lower-case letters to identify the discipline. I made a modest contribution to that Creation, and by a remarkable run of more good luck, I was able to parlay that modest contribution into a chair. Sometimes I am not too happy with all of this good luck because what emerged from that exercise was virtually an industry which like Topsy, has "just growed" without guidelines or evidence of careful thought.

I am reminded of Mr. Roberts, that wonderful novel of America's wartime Navy, in which Thomas Heggen writes disparagingly about the USS Reluctant: "It has shot down no enemy planes, nor has it fired upon any, nor has it seen any. It has sunk with its guns no enemy subs, but there was this once that it fired. This periscope, the lookout spotted it way off on the port beam and the Captain, who was scared almost out of his mind, gave the order: "Commence firing!" and the showing was really rather embarrassing...and all the time the unimpressed periscope stayed right there. [At closer range] it was identified as [a] branch, and didn't even look much like a periscope." I suspect that synthetic or stochastic or operational hydrology is much like the USS Reluctant; it has launched a few relatively successful sorties and a few successful careers. But it really has downed no enemy planes, nor has it sunk any enemy subs. What is worse, it has only rarely been fired in anger; it has engaged in lots of target practice, but rarely in combat. It has often been a bridesmaid, but rarely a bride.

I

A major difficulty with much of hydrologic science is that it is not a science at all. It is a black art, and not merely because it deals with black boxes. To be sure, there is a branch of hydrology which follows reasonably closely the rules of a conventional scientific discipline; I refer here to studies of some of the constituent elements of the hydrologic cycle, for example, infiltration, percolation, evaporation, transpiration, etc. One can visualize laboratory experiments which could be replicated and provide insight into underlying mechanism or which could provide multi-variate data on which to base serviceable empirical relationships until our understanding of causality displaces them. But when moving from the cool indifference of the laboratory to the noisy environment of even the most heavily instrumented field plots, the multiplicity of variables and the complexity of their relationships make it virtually impossible to develop from first principles reliable models of, and formulae for, the fundamental hydrologic fluxes. And if perchance such a relationship could be found, the richness of its parameterization would in most instances require an estimation algorithm which would mask the dependence of the estimates on identification of "the right model." There is likely to be enough noise in the observations so that fitting the parameters could be replaced by an intellectually less attractive, but numerically more efficient, strategy characterized by robust estimates of fewer parameters embedded in simpler—but incorrect—models. The question is whether one would rather be right or be President. As a scientist, I would rather be right; but as an engineer concerned with water-resource planning, I would rather be President. I would rather make designs (or decisions) which are a little bit wrong but which can survive the cumulative uncertainty—model, parameter, data, computational, demand, whatever—which the system and hence the model ultimately must face. The reward for being right comes from a different source; the penalty for being far wrong is immediate and severe, so the objective is to hedge the bet and cover the political and institutional tracks.

But leaving aside these scientific aspects of hydrology, the bulk of hydrologic analysis simply is not subject to verification in the conventional scientific manner; it cannot qualify as a conventional science. Let us explore some consequences of this in connection with applications to and of synthetic hydrology. Having been present at the Creation I can report absolutely authoritatively that synthetic hydrology was developed to identify weak links in large reservoir systems, and that it appeared heuristically sound to search for these potentially weak links by generating long traces of virgin hydrologic inputs which were routed through existing or proposed systems. It will be recalled that in the earliest application to the Clearwater River in Idaho there were no synthetic flood or drought hydrographs. Extreme monthly flow, candidate floods or droughts, were converted by reproducing one of the several characteristic hydrograph shapes taken at random from the historical record. Appropriate corrections were made to preserve the monthly flow volumes, which were the noble values. Our methodology for subsequent simulation of the Deleware River system was a little more sophisticated in that it reproduced the correlation between monthly volume and instantaneous peak, fitting the remainder of the hydrograph to preserve the noble flow volume and then to test the operating policy. But in these and similar early applications of synthetic hydrology, the principal purpose was to test relatively large systems for consistent failure of components under routine use.

This was also the motivation for multi-variate synthetic hydrology. We were interested in looking at larger and larger systems, with more hydrologic interconnections, and needed synthetic traces which preserved spatial and temporal correlations. Virtually all this early work focused on the multi-variate normal distribution because it was felt, although never proved, that most large systems would be relatively robust with respect to extremes, whereupon we probably could not learn very much about system performance from the extensive Monte Carlo analysis which would be required to reproduce extrema. These intuitive notions were subsequently verified by the extensive Monte Carlo experiments run by Wallis, Matalas, Slack and their colleagues at IBM and the USGS.

Thus our work at Havard de-emphasized the quest for "the right model" serviced by the "right parameters." In a very early study, I believe it was 1960 or 1961, Blair Bower and I thought it would be amusing to generate synthetic hydrology as if we did not know the population parameters but knew instead that they were estimated from relatively small samples drawn from known distributions. We replaced the point estimates of the moments by distributions of points, so that our random additive components in the standard synthetic hydrology formulation were added to random trend components. This gave us statistical noise built upon sampling noise and led, very early in the game, to a formalism we now know as regret analysis. In that particular study (of the Meramec basin) the minimax regret strategy called for using the normal distribution for all flows rather than its competitor, the lognormal, even though the lognormal clearly fit the data more closely than did the normal. So our design recommendations were based on the normal distribution, a result which was verified for floods under much less restrictive conditions by Slack, Matalas and Wallis some years later.

II

The next few years were characterized by the gradual acceptance of flow synthesis and the rapid advances in mathematical skills achieved by civil engineers—particularly those in the academic community. We became experts in stochastic processes and computer mathematics because these appeared to be useful and because systems analysis became a national passion. The Great Environmental Awakening of those years required new faculty appointments in water resources, typically within the civil engineering departments at our universities, and new appointments required demonstrations of mathematical fluency. So the national mania for systems methodology, coupled with available money, idle machines and the lure of tenure conspired to transform a modest statistical procedure into an industry. I do not believe we have been well served by the transformation.

We got untracked because we lost sight of the justification for synthetic hydrology. These can be listed as follows:

1. as a planning tool for large, inter-connected water-resource systems for which it is important to perform weak-link analysis based on unlikely combinations of hydrologic events;
2. exploration by massive Monte Carlo analysis of the sampling properties of moments and other parameters derived from hydrologic records;
3. sensitivity and regret analyses to adumbrate the importance of identifying the best distribution;
4. development, and testing of the sampling properties, of new indices for measuring basin characteristics and system response (e.g., indices of drought or water quality);
5. assessment of system operating rules under many combinations of routine hydrologic inflow; and
6. expression of the general view that exogenous variables affecting system performance (supply, demand, prices, etc.), the physical system and its model, and the parameters of that model are not known deterministically.

In addition to these six standard uses of synthetic hydrology, recent work has focused on the use of alternative traces of precipitation to drive flux and storage models of a catchment. But this emphasis was not always the case. Historically, concern for basin modeling was directed by civil engineers at the rainfall/runoff relationship because civil engineers are concerned primarily with hydrologic extremes—floods and droughts. Indeed, as synthetic hydrology became accepted by civil engineers, and as its use became standardized by widely available software packages, its original purposes became obfuscated and replaced by attention to extremes, particularly to floods, which would provide design requirements for reservoir systems. A favorite trick of mine in class is to present a flow record and then to argue, on the basis of synthetic hydrology generated from a defensible distribution and parameters, that an expensive flood control project is necessary. I then demonstrate how, for an appropriate consulting fee, the identical data can justify a distribution and parameters from which synthetic hydrology and a reasonable system objective would deny the need for flood protection. This kind of sophistry led to the need for Bulletin 17 (and 17A) in which it is dictated that for federal projects virtually all floods in the United States must be assumed to be derived from a specified distribution whose parameters are fit by a specified algorithm. While I take issue with the particular distribution which is chosen, and with the method of fitting its parameters, the notion of standardization is not naive even if Nature refuses to recognize the log-Pearson-3. The question is whether or not there is a better "national" distribution which might make better use of regional information in accordance with Bayesian inference and with the James-Stein paradox. I have no trouble in accepting the yoke imposed by a benevolent, robust, distribution function.

But if analysis of extremes is critical to the planning, design or operation of a project, it is a bootstrap operation to analyze the project by means of flood frequencies derived from synthetic flows. In the first place even if "the correct distribution" could be determined and verified, which it cannot, and if the correct parameters of that distribution could be found, which they cannot, running enormously long sequences of synthetic flows would routinely reproduce, with appropriate sampling error, the distribution that was originally provided. This should not be surprising. If short synthetic traces are used to test system response, the sampling instabilities of extrema will be manifest by a highly variable number of extremes generated in the several traces and, more importantly, by the inability of the algorithm appropriately to capture the effects of precipitation intensity and clustering of major storms. While these synthetic economic results might be useful in identifying instability of system performance, they certainly do not provide definitive answers to design questions. This leads to a curious situation. In many publications, the authors point to good agreement between the frequency of synthetic extrema and of observed extrema as evidence that they have identified "the right model." Such agreement does not warrant those conclusions, nor would lack of agreement justify the conclusion that the model is wrong. It fact, *all* models are wrong and the issue is only by how much. It is like the old joke—we have already established what you are, lady, but are now merely haggling about the price.

III.

Having thus disposed of what synthetic hydrology should not be trying to do, let us turn now to a modest success story by considering the relationship between synthetic hydrology and the Hurst phenomenon.

In the synthetic hydrologic activities associated with explaining the Hurst phenomenon, the amount of computation was, and continues to be legion. So it is gratifying that even though the Hurst phenomenon has not fully and satisfactorily been explained, at least significant serendipitous progress was made in related applications. Again, it all began quite innocently. Hurst and the great statistician, William Feller, gave the expected value and standard deviation of the range of partial sums for an experiment derived from a binominal coin-tossing game. We recall that (for development at 100% of the mean) the relationship between the expectation of the range and the record length involves the square root of N, the length of record. The coefficient of variation of the range of partial sums is 22% or 29%, depending on whether the adjusted sample mean or population mean is used in the calculation. But this simple result did not agree with the long sample of empirical evidence collected by Hurst, or with similar mass curve analyses. The agreement between observed and calculated ranges becomes acceptably close as the record length (N) becomes large; for N of about 75, the agreement generally is quite good.

There were many, myself included, who thought that more complicated flow models could produce synthetic records which, when subject to Rippl curve analysis, would provide exponents on the record length close to the empirical values of 0.72 and higher. It was generally agreed that no simple modification of the elementary lag-one Markovian model would suffice. This introduced that period during which complex models were introduced to "explain" the Hurst anomaly.

This is one of the most shameful periods of our hydrologic history. Multiple-lag models were introduced and justified on the basis of simple box-and-flux representations of basin hydrology. A variety of non-stationary models were proposed, along with long-memory models of fractional noise (Mandlebrot and Wallis) and a plethora of ARMA and ARIMA models. Box-Cox transforms were undertaken to convert non-normal distributions into their Gaussian counterparts, whereupon temporal generation was performed in Gaussian space and the inverse transform taken to revert to the original distribution (still an appealing approach, perhaps the best of the lot). Some authors compared alternative models, but most systematically ignored the fact that parameterization of any model introdued at least as much instability as the selection of an erroneous model in the first place, and that the test of a model's suitability (along with the algorithm for fitting its parameters) must lie in action space rather than parameter space.

What lasting results came of all this activity? Apart from the systematic exploration of a number of alternatives and formalisms for generating synthetic hydrology, not much was accomplished. It was demonstrated that a suitably complicated set of models would yield virtually any desired Hurst coefficient, but that some of these models were not hydrologically believable. Most models that showed some promise of explaining the Hurst phenomenon were characterized by relatively long memories; the Markov properties which made the earliest applications so neat and trim were lost. In fact, it was demonstrated that if a stream is formed by the confluence of two streams of lower order, and if each of the lower order streams is characterized by a Markovian process, then in general the higher order stream which they formed is not Markovian. Clearly the lagged serial correlation on the higher order stream is a function not only of the serial correlations of the confluent streams but also of their spatial relationship, and building this into the governing process for the downstream sum destroys the Markovian property, so the importance of Markov models was threatened.

It is particularly appealing to tune models on the basis of their ability to reproduce the Hurst coefficient. The coefficient is related to the range, which is a difference between extreme values of partial sums, and partial sums are related to integrals. It is well-known in numerical calculus that integration is a stabilizing process while differentiation is de-stabilizing. A numerical integral calculated by quadrature is generally less affected by instabilities in the data than in a numerical derivative, which can vary widely in magnitude and sign if a few of the data are greatly in error. The quadrature is far more robust, so that fitting functions and estimating parameters on the basis of a criterion using the Hurst coefficient is analogous to minimizing an integral loss function in action space. The hydrologic community should get high marks for its effort to resolve that problem.

In testing a number of alternative generating models for the effect of various lags and skew coefficients, we plotted a two- rather than one-dimensional Rippl diagram. (Incidentally, this year marks the one hundredth anniversary of the publication of Rippl's original paper). The conventional storage-yield diagram is one dimensional in that it represents a deterministic mapping between storage (or input) and yield (or output). The reliability with which that mapping obtains is not specified in most analyses because, following the prescription of Rippl, the function is based on the critical period of record and on the assumption that this critical sequence either will be reproduced during the economic lifetime of the structure or represents a sufficiently conservative basis for design. The two-dimensional generalization introduces an explicit level of reliability so that the storage-yield contours are Rippl diagrams, one for each level of system reliability. There are many ways of defining reliability (e.g., the probability of meeting the specified target, the total volume of water delivered divided by the nominal demand, an economic benefit or loss derived from a non-linear mapping or objective function, etc.). The principal advantage of this presentation is that it forces probabilistic considerations into the design process. Any two of the three parameters—storage, yield and reliability—uniquely define the third and the system reliability is considered right up front as a design parameter.

But perhaps equally important is the ability, after the fact, to plot actual capacities and projected yields for completed structures and to determine the level of reliability which the designers, wittingly or otherwise, inputed to their structures. Another way to think of such *post-hoc* analysis would be to plot the constructed storage and the announced reliability—if such a calculation were made—and to derive the imputed yield (typically expressed as a fraction of the mean annual flow or level of development). While all of this is appealing, in fact there are not many projects for which the requisite data are readily available. After the fact, when the design memoranda are filed, it generally is difficult to reconstruct the target yield or the specified level of reliability; perhaps there were several such values, or different uses, in a multi-attributed system. More common is the ability to draw a line at the design storage and note the trade-offs. Thus I could not find a large number of earlier results on which to base a retrospective study, but on the basis of perhaps 15 or 20 I was able to determine that very few designs actually stressed the available resource. That is, most specified high reliability and low yield (or level of development) *or* higher development but only modest reliability. The bulk of reservoirs constructed in the United States do not stress the available resource; theory shows that the storage requirement falls off very rapidly as the level of development and/or level of reliability recede from values near 100%, so that most storages perform well despite the fact that most were designed without explicit statistical consideration. This is verified by an informal comment made by the late Walter Langbein, who noted that as a national average, the level of development of storage in the United States is of the order of 60%. But as increasingly marginal sites become pressed into service, as most of the good sites become used, there will be a need to increase levels of development and target yields; this will move the design into that region of the contour map at which storage changes more rapidly, whereupon the system will become sensitive to changes in the design parameters.

IV.

The previous results while not profound, are at least interesting; nonetheless they hardly would justify all the fuss that has been made over synthetic hydrology. But there is some excitement in the old idea yet, and it pertains mostly to the use of synthetic hydrology for assessment of the nation's water resourcs. Congress requires a periodic assessment, and the first effort produced basically a comparison of supply and demand in each of several hundred water accounting units. Demand was projected by conventional estimating techniques, and supply was defined by the mean annual basin runoff or flow at the catchment outlet. In the second Assessment, the methodology was improved to accommodate some aspects of water quality, but the technique remains unsophisticated and generally inadequate.

The central feature of our most recent approach to assessment is that it is accomplished not merely by comparing surface and groundwater availability to the projected demand on a catchment by catchment basis, but that estimation of fluxes within the basin (i.e., among the storage components in the catchment) can be attached to values. Evapotranspiration and soil moisture are related to basin productivity, and runoff and soil moisture are related to the basin's capacity for development of municipal water supplies and assimilation of wastes. Thus we encounter another one of those paradoxes which encourages us to formulate and calibrate detailed models of basin physics to capture the basin's internal adjustments and to assign economic values thereto, while we are discouraged from doing this because most catchment studies provide very limited data. We seek a model of rainfall, runoff and adjustment of internal fluxes whose complexity lies between the very detailed and highly aggregated catchment models, and we seek a convenient and reliable way to estimate the parameters of such a model. It is the view of the Harvard group that assessments and catchment budgets are, for most practical purposes, indistinguishable.

We advocate that utility, and hence a measure of merit in an assessment, is related to soil moisture and evapotranspiration, as well as to the more customary measures of basin activity—precipitation and direct runnoff. Our work proposes a non-linear allocation of the sources, or total available water, to the major sinks. The total available water includes precipitation, soil moisture and free groundwater; the available sinks include the groundwater, direct runoff and losses by evapotranspiration. We propose a non-linear allocation among these three sinks, although portions of the allocation function can be approximated by linear rules. These simple proposals provide an important conceptual improvement over conventional rainfall/runoff relationships, most of which concentrate on one source or driving process (precipitation) and one sink or output process (direct runoff). It generally is true that direct runoff is only a small component of the total water budget for the catchment so that most conventional rainfall/runoff relationships do not closely fit the observations or cannot readily be generalized to other catchments or other (jackknifed) periods of the hydrologic record. The fits are typically highly site-specific.

Despite the shortcomings in most uses, identification of the rainfall/runoff relationship remains the central problem in surface water hydrology because precipitation and direct runoff generally are easily measured and characterized by long records at many stations, thereby providing a reliable source of data, and because direct runoff or streamflow at the basin outlet are the principal variables of interest in most catchment models and hydrologic assessments. This is because most models are directed at the analysis of hydrologic extrema, and because a reliable rainfall/runoff relationship provides a convenient mapping between the (frequently) long precipitation records and runoff events from which economic consequences and their probabilities can be identified. Then the resulting basin model and its parameters are the construct to be assessed because the model is in fact a surrogate for the real system and the parameters are its quantitative realization. Thus it follows that an important criterion for model selection is that it be parsimonious, or characterized by a small number of parameters.

Some years ago I proposed a simple box-and-arrow model as part of the physical justification for the long-lag models which might "explain" the Hurst phenomenon. The operating policy, or nature's way of allocating the available water within that model, was linear. More recently, Harold Thomas proposed a simple box model for the allocation of available water using a non-linear operating policy in which the allocation to the box called "soil moisture" depends on the propensity of the precipitation to infiltrate. The remainder of the precipitation is lost by evapotranspiration or runs off directly.

Because we fit models and parameters to observed time traces, the parameters (typically the moments of the available and measurable fluxes) become the noble values in the assessment procedures; this is an unfortunate result because the parameters themselves are artifacts of the fitting process and reflect potential instabilities of that process. As a consequence, we seek models and parameters which afford rational and stable descriptions of the catchment and which are responsive to anthropogenic change in the catchment. But more to the point, we seek models which contain stores and fluxes (for example, soil moisture and evapotranspiration) which are surrogates for the performance of the accounting unit in terms of its ability to support flora and fauna. They measure the *productivity* of the catchment, and can be mapped into assimilative capacity in order explicitly to assess water quality.

If such models and parameters can be found, if synthetic precipitation traces are provided, and if the anthropogenic changes contemplated for such a catchment could be reflected as changes in the parameters or in the routing of the water through the system, the synthetic traces then become a vehicle for introducing the statistical consequences of land-use change and development. Conventional models concentrate on precipitation and runoff—the hydrologic inputs and outputs—and thus their parameters generally are tuned to reproduce a close fit between observed and calculated values. If a model requires many parameters, and some in common use require more than 50 (although not all of these are equally important), there will be inadequate data unambiguously to define the parameters. Or perhaps the parameters are so brittle that trivial changes in the data base result in unacceptably large shifts in the parameters and hence in the inferences drawn from the assessment (and consequent decisions).

Even if a full data base is available, estimation of catchment parameters is potentially numerically sound. Evapotranspiration is the most significant sink for precipitation, accounting for some 64% of the precipitation over the land masses and about 110% of the precipitation over the oceans. Given that infiltration accounts for a significant fraction of the incident precipitation, direct runoff generally is a small residual. Being the difference between two large numbers of approximately the same magnitude, it is bound to be highly unstable and only weakly correlated with the incident precipitation. If we force the model to reproduce these unstable runoff values, it must be expected that, while the model can be tuned to perform well with respect to the observations, it cannot be expected that the parameters of the model will be robust. Thus our modelling effort concentrates on coupled formalisms for concurrently apportioning *all* the sources into *all* the sinks, even as we add that the underlying hydrologic processes are so complex that there is little hope of verifying any basin model within the near future. Thus dictates the need for empirical curve fitting along a frontier at which understanding is replaced by heuristic argument.

A problem with working along this frontier is that theorists, who have their roots in conventional science, generally have different objectives than the civil engineers who are interested primarily in the control of hydrologic extrema. These engineers typically bring few insights to basic hydrologic studies. Some of the insights developed in the most recent work, which we believe important for articulation of hydrologic basin models and for their conjunctive use with synthetic precipitation, are as follows:

1. The fluxes of the basin budget are analogous to the flows introduced by Simon Kuznets into his study of the national economic accounts. They include precipitation (income), direct runoff (export and consumption), evaporation (production), and soil moisture (cash balance).
2. Continuity is not necessarily obeyed deterministically in every time period, and some of the residual error introduced thereby is taken up in estimation of the parameters.
3. All the data required to fit the parameters unambiguously cannot be measured, so some must be fit by imposing continuity of *expected* fluxes over long time periods.

4. The available water is allocated by a non-linear operating rule to three sinks rather than the typical allocation only to runoff.
5. The operating rule is stochastic.
6. The parameters of the catchment model can be lumped parameters which reflect some physical significance or are merely artifacts of the fitting process.

Consider an expensive and extensive hypothetical study running for a period of several years in a densely-instrumented, experimental watershed. Suppose that precipitation, direct runoff, evapotranspiration and the soil moisture profile are measured. For each water year the hypothetical record is displayed in four columns. Earlier sections of this essay suggest that the evapotranspiration and soil moisture, two of the columns in the record, pertain to catchment production and hence are critical for basin assessment; unfortunately, they are the columns (or variables) most often unavailable. In fact, suppose that these two columns are erased. The problem is now whether a reliable reconstruction of these columns can be made using only the information in the first two columns—rainfall and runoff. In this form, the question was posed by my colleague, Harold A. Thomas, in our recent study for the U.S. Geological Survey. He wrote: "The task might be viewed perhaps as an extreme case of using statistical techniques for filling in missing data. However, such techniques have been devised primarily for patchwork and not for major augmentation of the data base. Moreover, solution by conventional techniques of multiple linear regression does not appear to be feasible... One might conjecture that solution has been deemed impossible. However, it will be shown that reconstruction is possible though tedious and inexact in some cases."

He proposes that because the total through-put of the system is much larger than annual variations in carry-over storage, the cumulative evapotranspiration over several years must be closely balanced by the difference between cumulative rainfall and runoff. The difference between them, the cumulative variation in the carry-over storage, is so small as to be negligible. The opposite is true for arid zones, suggesting that the model might not fit so well in these cases.

The four time series are related by six cross-correlation coefficients and four serial correlation coefficients, or ten in all. In the truncated data set which consists only of rainfall and runoff data there are only three coefficients; successful reconstruction of the complete data base therefore requires that seven missing correlation coefficients must be estimated. In addition, the moments of the evaporation and storage time series must be reproduced. As stated earlier, the synthesis of carbohydrates and other organic compounds, which are surrogates for primary production within the catchment, is more closely correlated with transpiration than with rainfall or runoff. Thus ecological activity within the community is measured by transpiration and evaporation, which is physically acceptable because these measure the interception by plants. However, this evaporation takes place from interception on surfaces above the ground, and might reduce the loss of soil moisture. Hence the evaporation and residual soil moisture profile are related, and a successful catchment model must reflect this potential for substitution by the plant. The effects of successive dry years can be assessed more accurately by deviations from normal levels of the joint value of evaporation and residual soil moisture than by scalar shortfalls from either precipitation or runoff. This supports the earlier contention that, as stated by Thomas, "...the augmented hydrological data could make possible improved classification of basins in different regions, and quantification of their carrying capacity for different types of land use."

The model is designed for annual events and carry-over storage. In one of our applications, to the irrigation scheme on the People's Victory Canal in China, we attempted to fit the model to seasonal and monthly values. Although it is not entirely fair to evaluate the performance of the model in this context because of the paucity of data, it was clear to us that the model performs very much better when applied to annual events, for which it can be assumed that changes in soil moisture are small relative to basin throughput.

The model assumes that annual precipitation is independent of the other variables in the system. Thomas proposed two versions of the model. In the first, all the variables which comprise the accounting, and which together represent the set of sinks to which the basin allocates the total available water, are assumed to be normally distributed. In the second, these same variables are taken to be lognormal. At the moment, while we have made several trial runs with the lognormal version, most of our experience is with the normal distribution. An important result is that while the individual values of the parameters vary considerably, the parameters are interdependent so their joint variation might be compensatory and lead to approximately equivalent consequences when used as part of an assessment algorithm—e.g., in evaluating the production of the biomass in a catchment, the assimilative capacity of the resource, or other economic use of the basin.

The available water in the basin is a major determinant of the fractions that are allocated to runoff, evapotranspiration and carry-over basin storage at the end of the water year. In fact, the allocation depends on energy considerations and rainfall intensity (the spacing, number and distribution of storms). In general, the model divides the total available water into runoff, evapotranspiration and residual storage levels associated with a year of normal rainfall intensity and into perturbations of these allocations; the perturbations are assumed to have expectations of zero and to depend on the intensity, a surrogate for energy. The effect of variations in intensity on carry-over storage will usually be much smaller than their effect on runoff or evapotranspiration. In other words, the observed runoff is conveniently thought to originate from two components; the first corresponds to the effect of the precipitation volume and of the antecedent storage while the second accounts for other sources of deviation (the peakedness and timing of storms, stochastic fluctuations in potential evapotranspiration, etc.). The model assumes further that these two effects are statistically independent, and that similar divisions can be made for the annual evapotranspiration and the carry-over storage.

This is not the proper forum for presenting the details of the Thomas model, nor is it my desire to display my colleague's work before he offers it to the profession over his signature. Instead, I wanted merely to sketch the outline of his proposal and some of the insights it has provided.

Thomas makes a very important point here: "In most basin waters in the deep subsoil and in fractured rock is hydrologically and biologically inactive and there is no physically identifiable datum for which the storage is zero. However, the magnitude of the variation in storage about the mean level from year to year is a strong correlate of biological productivity and stability. Therefore estimates of the standard deviation of storage, together with those of evapotranspiration, are important results of the analysis." In other words, it is not necessary to define the mean storage but the movements about that mean level are critical; these migrations define the nature and level of biological activity which can be supported and hence serve as surrogates for basin assessment. The details of the iterations required to reproduce the missing data columns are not important for the moment; they are in our report to the U.S. Geological Survey and ultimately will appear in the journals. The important point is that we made a large number of trial runs using synthetically generated precipitation values and tested the ability of the model to reproduce the correlation coefficients and the missing columns. In all instances such constructions proved successful.

V.

What now of synthetic traces? It seems clear that some of the assumptions of normality and independence implicit in the analytical formulation given by Thomas can be relaxed and a variety of models imposed on the precipitation sequence to test the performance of the basin under a variety of alternative virgin flow regimes and, more importantly from the point of view of assessment, under a variety of anthropogenic basin changes. For example, if it is planned to pave or clearcut or develop or otherwise change part of the basin, it is reasonable to ask how these changes might affect the several mean values and standard deviations and then to inquire how these might alter the generated (filled-in) values of evaporation and soil moisture—the productive or business end of the record. If such changes are significant, then we have a way unambiguously to map basin development by an ecologically sensitive ranking function, and unambiguously to determine, by calculation of various regret matrices, whether it pays to seek more complex models to represent the precipitation and flow regimes. The basin is completely characterized by a relatively small number of parameters, somewhere between the three or four parameters suggested by simplistic linear models and the many parameters required by such models as the Stanford Watershed Model and its descendants. Effectively, synthetic hydrology can be rejuvenated because it produces not only synthetic streamflow, but more nearly what was originally promised in the use of the word "hydrology:" that is, a more substantive and meaningful, a more complete and comprehensive, vector description of the meteorological processes which impact the basin and which ultimately affect not only the design of storage systems but the entire planning process for use of the basin as a resource.

VI.

The fugue is now complete. We passed from simplistic lag-one and Markovian models of streamflow through a plethora of complicated mathematical exercises to describe natural phenomena, back to a more fundamental, causal set of relationships which can generally be confirmed by observation. On reflection, this process is neither unexpected nor undesirable. It seems sensible to develop mathematical and statistical insights, to attempt to detect their limitations, and then to step back a bit and ask what we can do with our more powerful tools. Thus I would like to think that not only was I present at the first Creation, but that I participated in the Re-Creation of that hydrologic inquiry which confers a higher use of some of those techniques developed in one context but which now find application elsewhere. This ultimately is the highest goal of science, and one on which I am sure Chester Kisiel, Nicholas Matalas (who preceded me on this podium) and some of the distinguished people who will follow me, would certainly agree.

MYRON B FIERING
Harvard University, Cambridge, MA 02138

Myron B. Fiering was born in New York City where he had his primary and secondary education, preparing for college at the Bronx School of Science. After taking an undergraduate degree in engineering sciences at Harvard College, Fiering practiced civil engineering and travelled extensively for the New York consulting firm of Tippetts-Abbott-McCarthy-Stratton. Two years later, he returned to Harvard's Graduate School and from 1957 to 1960 participated in the Harvard Water Program which combined hydrology, engineering, statistics, computation, and social sciences in innovative and exciting ways.

Following a brief stay of one year at UCLA in 1960, Fiering returned to Harvard, and has been on the faculty of the Division of Applied Sciences since 1961. Soon after this initial appointment, Fiering became Gordon McKay Professor of Engineering and Applied Mathematics, bringing computers and statistical modelling into water resource planning, maintaining continuously a leadership role in the application of operations research methods to problems of resource allocation.

Recently, Fiering initiated and was chairman of a study group on the Scientific Basis of Water Resource Management, which was sponsored by the Geophysics Research Board of the National Academy of Sciences. The findings of this study group were published by NAS in 1982.

FOREWORD

The basic problem in water resources is that of water management. Specifically, the crucial issue is to reconcile the discrepancy between the flux observed in the hydrological cycle and a desired flux, so that certain societal objectives may be attained. This "reconciliation" has a number of dimensions, the most important of them being timing, geographic distribution, quality of water, and the institutional framework which may either facilitate or inhibit the resolution of this issue. A related question is that of the response of physical hydrologic systems (as well as chemical, biological, or socio-economic hydrologic systems) to management stresses. An example of a rather extreme management stress is the mining of groundwater.

Management includes collection and processing of information. Following this activity it is assumed that managers have a better understanding of the physical (and of the chemical and biological) and socio-economic processes with which they have to deal. Having this better understanding, it is expected that managers will make wiser decisions in relation to the societal objectives which we—the society—deem necessary to be attained. These decisions—wise or otherwise—are increasingly difficult to make because there are more water users competing for a virtually fixed amount of water available at a given cost in a region and, in addition, the demand of certain user categories exhibits an upward trend. The increase in population and the steady rise in our standard of living, which is a remarkable and laudable feat in itself, places an enormous demand burden on regional water resources, particularly in the arid Southwest. In some communities, we are not satisfied only with the indoor comforts provided by appliances such as laundry machines and dishwashers which are heavy users of water, but expect also a well-watered lawn, lush landscaping of the outdoors, extensive golf courses, and artificial lakes in which to boat, swim, and fish. Thus, water scarcity becomes a fact of life and a political issue.

When faced with scarcity—quantitative or quality-induced—the political system tends to produce regulations. Given the great hydrologic diversity of the United States, can the "regulator" and the "manager" be one and the same person under all conditions? Under any condition? Given this diversity, how can regional water resources be managed efficiently and equitably? Can we hope for some "optimal" management policies? these are some of the crucial issues addressed by the Third Chester C. Kisiel Memorial Lecture.

Nathan Buras

WATER MANAGEMENT IN THE UNITED STATES—A DEMOCRATIC PROCESS (WHO ARE THE MANAGERS?)

J. D. Bredehoeft

INTRODUCTION

Water use in much of the western part of the United States has reached a state where virtually all the available water is currently fully allocated and utilized. This is especially true in the Southwest where in Southern California and Arizona shortages of surface water—streamflow—are made up by the mining of ground water. In Arizona, ground water is currently being mined at a rate of approximately 2 to 2½ million acre feet per year.

Given the current state of affairs, the arid and semi-arid West in particular, the Southwest is not "out" of water rather all the water available—which is a substantial quantity—is being fully used. This of course means that as we as a society consider new activities which require water, the water will have to be diverted from ongoing activities. Since irrigated agriculture is by far the dominant user of water in the West, consuming approximately 90 percent of the supply, water will continue to be shifted from agriculture to other municipal and industrial uses. Water is a limiting commodity over much of the West, in that sense there is a shortage of water. This does not mean that there is any less water available, only that the competition for what is available has increased (Bredehoeft, 1984). In a classic economic sense, one would expect the price of water to increase; however, commonly water is not allocated by allowing the marketplace to operate freely.

Given the increased competition for water, there is an accompanying emphasis on the part of hydrologists for "better" water management. We as hydrologists have been investigating problems associated with water management with increased vigor during the past 20 years or so. With the advent of modern digital computers it has been possible to simulate hydrologic systems and examine their response to various stresses. By examining the system response we have been attempting to provide insights into how to better manage our water resources. As our computer resources have grown, so has our sophistication in simulating water resources systems. We have taken the concepts of operations research and applied them to the analysis of water resources systems. We have worried about such ideas as the global optimum—had we achieved it?—and other such equally sophisticated, and perhaps esoteric, questions.

Within the U.S. Geological Survey, we have concerned ourselves largely with the hydrologic question of how the physical (and perhaps chemical) hydrologic systems will respond to stresses. We developed computer simulation codes to analyze such problems. Over the past 10 to 15 years we have probably simulated—modeled—several hundred such systems, mostly ground water systems. Many of these model studies were financed jointly by state and local governmental agencies—half the money supplied locally and half by the Federal Government. In many, if not most instances, the studies were promoted and sold with the idea that they would provide for better day-to-day management of the resource. However, in retrospect almost none of the model analyses were utilized for any kind of direct resource management.

Given the experience of the USGS where for all practical purposes very few of our model analyses were used in direct management, it behooves us to reflect on why this is the case. (I do not believe the experience of the USGS with respect to water is very different from that of other scientific investigative groups in this country.) It is the purpose of this paper to examine how water is managed in the United States and then to reflect on what these insights in water management suggest for the role of hydrology in our society.

WATER MANAGEMENT: SOME COMMENTS

Our concept of hydrology is based upon the Hydrologic Cycle. Water within the cycle is continually flowing. Driven by the energy of the sun it moves from the ocean to clouds where it is transported in the atmosphere, falls as rain on the land, is used by plants and transpired back to the atmosphere, or runs off as streams which ultimately bring the water back to the oceans. The Hydrologic Cycle for the conterminous United States is depicted in Figure 1 (U.S. Water Resources Council, 1978).

For the conterminous United States the water use by man, defined either in terms of what we withdraw from the system or what we consume (meaning what we convert by evaporation or transpiration into vapor and put back into the atmosphere), is a very small fraction of either what runs back as streamflow to the ocean or an even smaller fraction of water which falls as rainfall. Of course the problem is that flux through the Hydrologic Cycle is not distributed equally in either time or space. This unequal distribution of water in both time and space gives rise to water management.

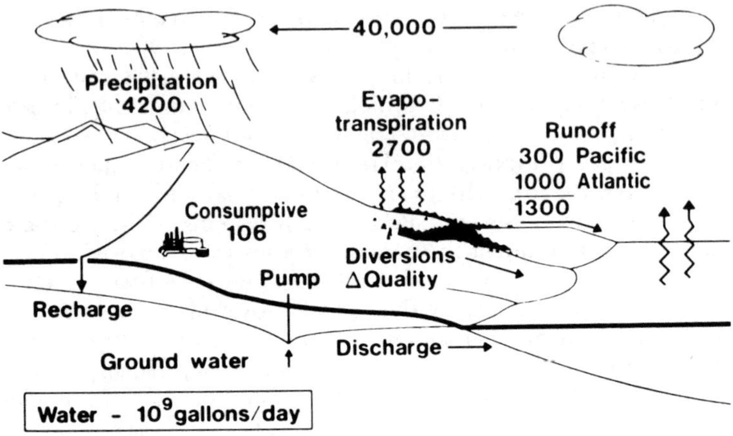

Figure 1. Schematic diagram of the Hydrologic Cycle for the conterminous United States.

Water management in concept is really quite simple with only two or perhaps three possible actions:

1. Management can change the timing of the flows through the Hydrologic Cycle.
2. Management can change the geographic distribution of flow through the system.
3. Management can mine ground water.

These actions have a possible side effect of changing the water quality; a fourth action within the context of management is:

4. Management can change the quality of the water flowing in the cycle.

The actions are obvious, but perhaps a few examples are in order. In the West much of the precipitation comes as snowfall on the mountain ranges during winter months. This snowfall melts and runs off during the late spring and early summer in most places. However, the distribution of runoff is not in phase with the demand for water of most crops grown in the West. The peak in crop water demand is later, in early to mid-summer, than the runoff which peaks in late spring.

One of the primary reasons for the building of reservoirs for irrigation supplies in the West is to bring the supply more nearly in phase with the demand. The seasonal crop requirement function along with the sources of supply of both ground and surface water for the South Platte in Colorado are depicted schematically in Figure 2.

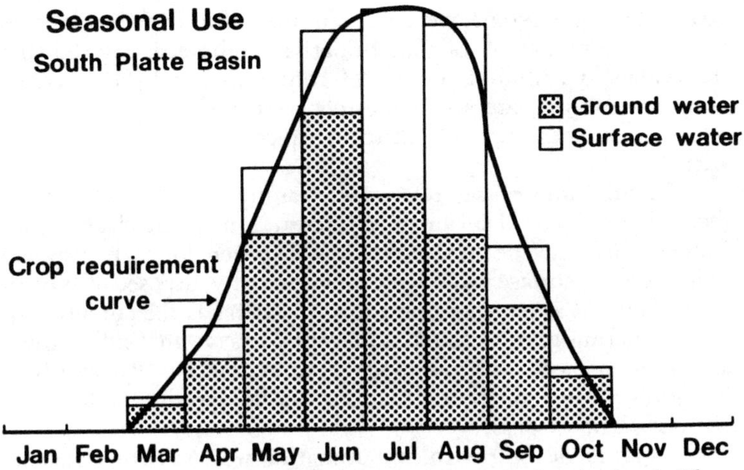

Figure 2. Crop requirement and sources of supply for the South Platte Valley in Colorado. The surface water available does not meet crop demand.

California is a prime example of the redistribution of water through water management. Water is moved from the Sacramento River drainage in northern California 400 or 500 miles south to Los Angeles. The major interbasin transfers in California exceed several million acre-feet of water annually (State of California 1979).

Various areas in the country are dependent upon mining ground water for a major portion of their water supply. The principal areas are the High Plains of New Mexico, Texas, Oklahoma, and to some extent, Kansas and Colorado; Central Arizona; and Southern California (Lucky, et al., 1981). In these areas various aquifers were filled with water over geologic time. Man is now removing water from these systems at rates which are much higher than their rate of replenishment, in effect mining water from these systems. Man has mined many of the earth's natural resources. In some sense, mining water is not very different from mining other minerals or fossil fuels; many of the same societal value judgments are involved.

One of our growing concerns has been a recognition that our management actions have the potential to change water quality. Indeed, in recent years we have moved to decrease the degradation in water quality caused by various point sources of pollution. As a nation, we passed a Clean Water Act which was designed to clean up the nation's rivers. We have spent billions in the pursuit of this goal and to some extent, we have been successful. In our attempt to clean up point sources of contamination we have discovered that the widely distributed sources of contamination, the so-called nonpoint sources, are also major contributors of pollution. The nonpoint sources have proven to be much more difficult to deal with.

In this country our efforts at cleaning up the environment focused initially on clean air and clean water (meaning clean surface water). This focus meant that wastes which were formerly disposed of in the air or surface water were now generally disposed of as on or in the land. Waste disposal in the solid earth has the potential for contaminating ground water. In recent years we are finding more and more instances of ground-water contamination (Pye and Kelley, 1984). Many of these situations can be traced to faulty practices of waste disposal.

In issues dealing with contamination we, as scientists, commonly focus on the changes in water chemistry—water quality. However, our real concerns are what impact do these changes in water chemistry have on two ecosystems which concern us: mankind itself, and the existing "natural" ecology. The interaction between changes in water chemistry and its impact on either man or the existing natural ecosystem is not well understood. While this is an area of active research, for the moment we make judgments based upon a surrogate, changes in water chemistry, for our real concerns—impacts on the ecosystem.

NATURE OF WATER MANAGEMENT

Water management, especially on a large scale, is a process of continuous planning. One can picture continuous planning and development as illustrated in Figure 3. We are continually concerned with implementing the next increment of additional development.

In most instances, the early development does not come anywhere near the total capacity of the system. Usually management decisions in the early stages of development, when there is plenty of water, are not nearly so critical as decisions once the system is close to full development.

There is an infinity of possible paths to full development of a particular water resources system. Two extreme paths are indicated on Figure 3. The lower path is one in which development gradually increases until it reaches a level of full development, at no point do we exceed full development. The other extreme, illustrated by the upper path, is a development which at its peak exceeds full development and then declines downward to a level which can be

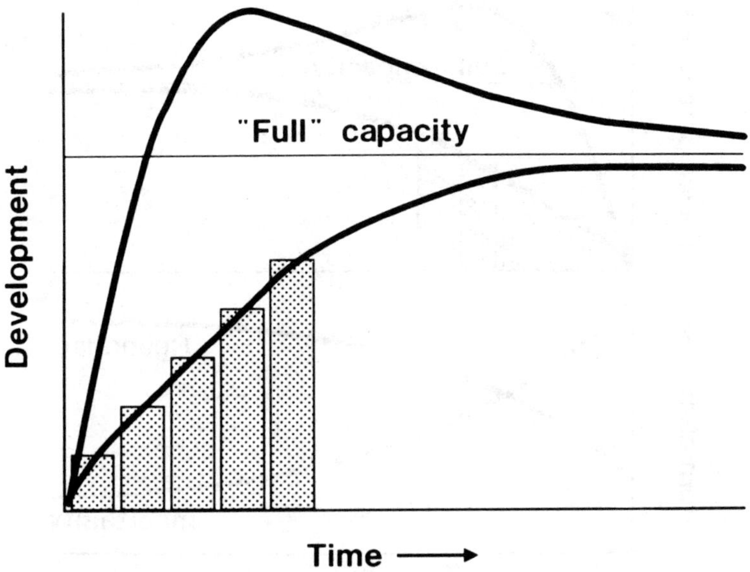

Figure 3. Schematic representation of the process of continuous planning and development. Each stipled vertical bar represents implementation of an increment of additional development.

sustained through time. The second path, which exceeds "full" development, depends upon storage in the system which can be exploited or on a series of unusual hydrologic conditions, such as a sequence of very wet years. It is not unusual in ground-water systems to have a development which follows the upper path and exceeds a level at which it can be sustained indefinitely.

There seems to be a tendency to judge a development path which exceeds full development for some period as undesirable. One can show examples of development which follow either path; obviously each is viable. It becomes a matter of value judgment as to which policy society chooses. In our society such choices are commonly made by our political institutions. One can contrast ground-water development in Arizona with that in Nevada. Arizona's development has proceeded to the point where currently ground water is being mined at a rate of approximately two to two and a half million acre feet annually. Nevada, on the other hand, has had a policy of restricting development to a level that can be sustained indefinitely. Arizona has moved through recent ground-water legislation to curtail ground-water mining with the objective of eliminating it entirely sometime after the year 2000.

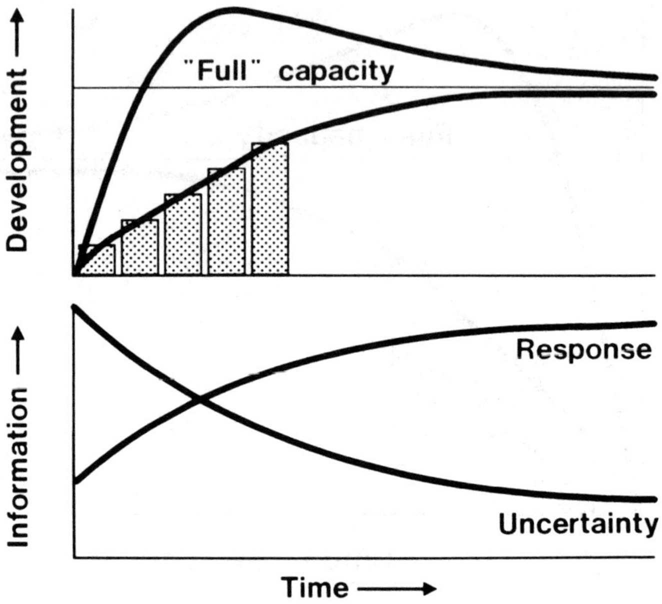

Figure 4. The role of information in reducing uncertainty during the process of development.

One can speculate to what extent different water policies in various western states have either impeded or enhanced economic development.

One important management action, often overlooked, is the collection of information during the period of development. The role of information is illustrated in Figure 4.

As a system is developed, information is collected on how the system responds. For example, in a ground-water development as pumping begins, water levels in the aquifer decline. They may stabilize with time, or continue to decline. Data on both the rates of pumping and the water levels in the aquifer are important to any future analysis of the system and to management decisions. Intelligent water management anticipates future problems and begins a program of systematically collecting information long before a crisis arises.

As information is collected about how a particular system responds to stress, uncertainty in predicting the future response of the system is reduced. In many large capacity water resources systems it is usual to reach full capacity following many years of

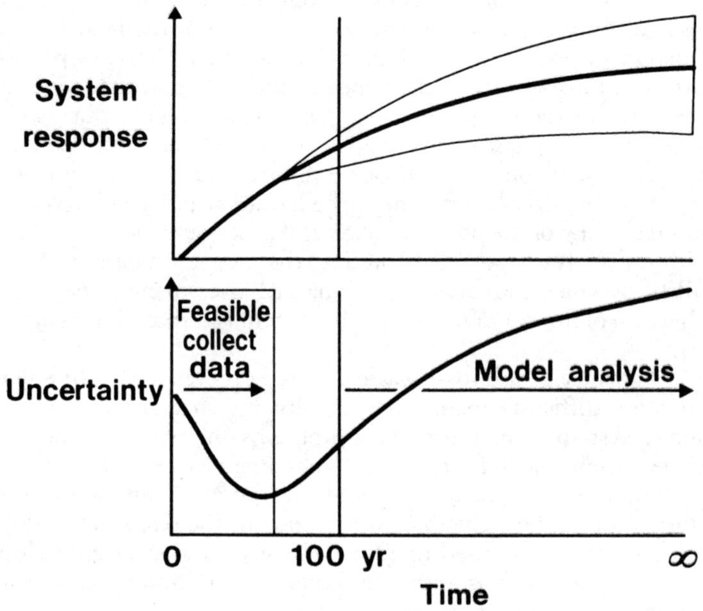

Figure 5. The long-time frame problem for the hydrologist.

development. Hopefully, critical information which describes how the system responded to development was collected during the period of development. It is this information base which makes possible analysis of the system, and more intelligent management.

There are several critical environmental problems in which the time frame of the problem is such that it is virtually impossible to collect meaningful data describing the response of the system before management decisions which are virtually irreversible are made. Two such problems are:

1. Toxic waste disposal, both nuclear and other toxic wastes.
2. Carbon dioxide build-up in the atmosphere (National Research Council, 1977).

Both these problems require management decisions in the relatively near future. In either case, the data will be insufficient to provide unambiguous information on how the system is responding. Decisions will have to be made on our scientific understanding of the processes involved with little or no data on the dynamics of how the system responds.

The problem of predicting the long-time response of a system with little or no meaningful response data is depicted in Figure 5.

In the case of a nuclear repository the most probable mechanism of contaminant transport away from the repository is transport by ground water. This will be an extremely slow process with the probable path back to the biosphere involving thousands, if not tens of thousands, of years. Observations made during the probable active phase of the repository, several decades to perhaps 100 years, will contain little or no information pertaining to the long-term hazard. Assessment of the hazard will depend upon our understanding of the physical, chemical processes involved and our ability to analyze (model, if you like) the long-term hazard. There will be no chance for refining our hazard assessment through the collection of further data on how the system is responding dynamically.

The nuclear and toxic waste disposal problems pose different, and more difficult, management problems than most water resources systems. For most water problems the planning and implementation usually proceed in increments over a number of decades, often a century or more. As pointed out above, the information on how the system responds to the stress of development, is commonly used to adjust and scale continuing development. It is for this reason that hydrologists are almost continously involved in analysing the same system.

Within the U.S. Geological Survey we haved had an office on Long Island which has been involved in studying ground-water problems of the island for more than 40 years. One might imagine that after 40 years, the USGS had investigated all the relevant and interesting ground-water problems of Long Island. However, the aquifers on the island are still being heavily utilized both for water supply and waste disposal. Management decisions are being continuously made. These decisions require new analyses as well as additional information. I am confident this situation will continue as long as we as a society attempt to utilize the system.

THE WATER INDUSTRY

As hydrologists, we commonly discuss water management in terms of "the water manager." In recent years, hydrologists have attempted to adapt many of the techniques of operations research and systems engineering to the management of water resources. Implicit in much of the methodology of operations research, especially the optimization techniques, is the assumption that there exists a "manager" sufficiently powerful to implement an optimal scheme of operation or development. I think it is useful for those of us involved in both hydrology as well as water resources management to ask ourselves: "Does such an all powerful water manager exist?" In a broader context I think it is useful to examine how water is managed in this country.

As hydrologists, we generally tend to think that the various water managers with whom we are associated have more power to manage water than their legal authority gives them. For the past several years I have been responsible for the water resources activities of the U.S. Geological Survey in the eight western states. In this position I have had the opportunity to talk to a number of the more important "water managers" in the West. I have been continually struck by the extent their management options are constrained by their legal authority. The individual most hydrologists in the West would identify as the "water manager" is the State Engineer or his counterpart. After discussing with a number of Western State Engineers their role, I believe it is better described as a *regulator* rather than a manager.

As this perception was reinforced by various people in positions of authority, it seemed increasingly interesting to look at the entire structure of how we manage water. Certainly there is no question that water is highly managed, especially in the western United States.

In a state like California there are a myriad of involved parties each playing a role in water management—local water companies, local water districts, the State Department of Water Resources, the State Water Control Board, the U.S. Bureau of Reclamation, the U.S. Corps of Engineers, the U.S. Environmental Protection Agency, the state courts, the Federal courts—the list goes on and on. Looking at this complex list of interested and involved parties we can ask: "Is there some pattern in the role each plays?"

One of the more perceptive works on water management is the book *Northern California's Water Industry* by Bain. Caves, and Margolis (1966). Bain et al. (1966) examined the distribution of water much as one would examine any industry in our society. That approach is particularly enlightening in defining the roles of the many parties involved, and has strongly influenced my ideas.

I have attempted to abstract the roles of the principal parties involved in Table 1. I have broken the industry into three general groups: users (or consumers); suppliers, and regulators. Each group is generally represented at the Federal, state, and local level. There is an unusually large number of regulators which I have also broken into three groups: capital investors, direct regulators, and the courts. It is of interest to examine what are the principal motivators for the various entities within the water industry.

I will try to make a few brief comments on the principal motivations for most of the entities identified in Table 1. It is

TABLE 1

Structure of Water Industry

Users	Capital Suppliers	Direct Investors	Regulators	Law and Courts
	Federal			
	Bureau of Reclamation	U.S. Congress	FERC	Federal
	Corps of Engineers		EPA	
	State	Legislators	State Eng.	State
			State EPA	
Municipal and Industrial	Regional	Public bonds and taxes	Indian tribes	"Water Courts"
Power Company	Local			
"Indian Tribes"		Concerned interest groups		
Individuals	Individuals (well owners)			

interesting to examine the motives of each of the involved parties with the question in mind: "Can an industry, structured as the water industry is, be 'managed' in some kind of 'optimal' manner?"

LAW AND THE COURTS

Water is the subject of continued legislation both at the state and national levels. Traditionally the states have established laws which allocate the available "quantity" of surface and ground water. In some instances state constitutions establish a hierarchy of beneficial water use. Within the last two decades environmental concerns of the nation have resulted in the Federal Environmental Protection Agency along with Federal legislation which is designed to clean up the natural environment of the entire nation. National environmental legislation has thrust the Federal government into local water management, an area that has been traditionally a state prerogative. Both Federal and state legislative bodies are involved in two key regulatory activities with respect to water:

1. creating water law and
2. funding water projects.

A number of authors going back to the last century have pointed out that water in the arid West makes the land valuable, while this is perhaps an overstatement certainly water is the critical commodity to western agriculture, as it may be to urban and industrial growth in the West. Water has generally been appropriated over most of the western United States by the doctrine of prior appropriation. Traditionally our legal system has moved to protect an individual's property. It is fair to say that our legal system has generally moved to protect the water rights of individual users, by so doing they have protected the property values of individuals.

On the major rivers of the West a series of interstate compacts have been negotiated through the Federal courts. These compacts define the allocation of water between states and restrict interstate competition for water.

Along many of the rivers of the West ground water and surface water are inimately tied together by the geology of the physical system. This situation exists along the Platte River in Colorado, Nebraska, and Wyoming, along the Arkansas River in Colorado, and along the Rio Grande River in Colorado and New Mexico, to name a few examples. The development of ground water in these systems impacts the available surface water; while the impacts are often delayed, ultimately the streamflow is depleted at some time. However, if operated effectively the ground-water system can be used as a reservoir to supplement the total water available during the growing season. In many such systems utilization of both ground water and surface water has increased the agricultural output of the total system dramatically. This is the concept of conjunctive use of both ground and surface water.

Conjunctive use has been a particularly difficult institutional problem for both legislation and regulation. Senior surface water rights are damaged by ground-water pumping in these systems. Conversely the economic output of the entire system is greatly enhanced by conjunctive use. The problem has been to find damages to senior surface water rights. Perhaps the most interesting solution is the one reached in Colorado in which the state requires the ground-water pumpers to create a cooperative and collectively augment stream flow during periods in which pumping has adverse impacts.

While it is dangerous to generalize, there have been attempts in various states to tie water more closely to the land. These attempts seem to be motivated by members of the farming community with aspirations of preserving an agrarian society. Tying water to land certainly impedes a "market" for water. Interestingly, many of the recent attempts to tie water and land together have been defeated by the "owners" of water rights who recognize that the right to use water may be more valuable if not attached to the land. This reflects the present situation in which most if not all the water available is fully utilized. Urban and industrial growth in many places can only come with a shift of water resources away from agriculture. Recently urban and industrial users have been willing to pay handsome prices to acquire water rights.

THE DIRECT REGULATORS

In order to undertake construction of any sort on a navigable stream in this country one must have permits from the Federal Corps of Engineers and the Federal Energy Regulatory Commission (FERC). In recent years it has licensed numerous low-head, hydroelectric projects throughout the country. In some instances the FERC has required additional monitoring to be undertaken, usually of stream flow.

The Corps of Engineers, under a statute dating back to the late 1800's, has a responsibility for restricting adverse impacts to the navigable streams of the entire country. Recently these impacts have been extended to include water quality, including the impacts of sediment and bank erosion. The Corps has thus become a major regulatory agency, granting and denying permits for almost any action, large and small, on the navigable streams of the country. Most actions involve some assessment of environmental impacts on which the Corps passes judgment.

The Federal Environmental Protection Agency through legislation involved in the several Clean Water Acts, and more recently statutes involving the disposition of toxic wastes and protection of ground water, has regulatory responsibility involving water quality. In many instances, Federal legislation has provided an incentive for the states to act to protect water quality. Most of the Federal legislation has been designed to pass the responsibility for action to the states. However, in those instances in which the states have been reluctant to act the Federal Government through EPA has stepped in with direct regulatory authority. Legislation associated with environmental protection has threatened the traditional role of state primacy with respect to managing their own water resources.

As suggested above if one asked the involved citizen or hydrologist who was the "water manager" in most western states, most would respond the State Engineer. Over the past 30 or 40 years there has been a distinguished group of western state engineers. Many of them have been dedicated civil servants who have greatly influenced western water policy. I have had the opportunity to talk with many of these individuals. In discussing water management with them I have been struck with their role of implementing state law. Their role, I believe, is one of regulatory authority rather than a manager in any traditional sense.

An extension of the State Engineer's Office is often the local water-master who actually allocates the streamflow available in a given river basin within a state. He administers the allocation of water according to the existing water rights. This individual usually exercises considerable judgment in how he allocates the available supply. I think it is surprising that there is as little controversy as there is with the administration of the local water-masters. This is perhaps a tribute to the individuals selected for these positions.

In recent years the states have generally had a state water quality agency which parallels the Federal EPA. These state agencies have worked very closely with the Federal EPA. They have implemented local environmental statutes mandated by Federal legislation. In most cases they have been the recipient of large Federal funds directed to controlling problems of environmental quality; one of the major concerns has been stream quality. Recently EPA is struggling to control ground-water contamination, and has recently isued a "ground-water protection strategy."

I believe it is a fair criticism of the Federal EPA that its creation did not take into account that water quality is intimately linked to water quantity. Because quantity and quality are intimately linked they must be managed together. A number of states, with some notable exceptions, have created state agencies which generally reflect the Federal Government structure and in so doing have maintained different agencies regulating quality from quantity. This separation of regulatory functions will, I believe, be a nagging problem in the management of water resources in many states.

The environmental concern of the nation focused first on air and stream (surface water) quality. Obviously if one wishes to dispose of wastes, they can generally be disposed of as a gas, a liquid, or a solid. Stated another way, we can put wastes into the air, the surface water, or the solid earth. In cleaning up the air and surface water we pushed society's wastes toward disposal in the solid earth. We are beginning to realize that some of our disposal practices have contaminated significant quantities of ground water, some of which we are dependent upon for water supply (National ResearchCouncil, 1984). For a modern, technologically-based society, waste disposal is a major problem, a problem which is growing. Since our water supplies are often what is impacted most by our waste disposal practices, these problems inevitably involve problems of water management.

In recent years western Indian tribes have become involved in regulating water. Along the west coast of the United States the Indians have recently been awarded by the courts one-half of the salmon population for their exclusive use. Since the salmon move through and spawn in many of, if not most of, the streams of the West, the Indians have now a legitimate concern in the salmon habitat. This has made the Indians potential players in virtually all water management decisions involving surface water in many of the western states. This decision is too recent to know to what extent the Indians will insist on participation in regulating water management. Early indications from Washington and Oregon suggest that the various Indian tribes expect to be heavily involved.

In the arid and semiarid areas of the West the Indians have raised the issue of how much water they are entitled to under the conditions stated in their original treaties. Various tribes are laying claim to large quantities of water already fully utilized. These claims have the potential of upsetting the existing right structure. The Indian claims become more valuable as the competition for existing supplies increase; the Indians do not seem to be in any hurry to settle their claims. This leaves a degree of uncertainty, especially for the traditional regulatory agencies.

CAPITAL INVESTORS

Large water projects require large capital investments. It is the general policy in this country that major projects involving nagivation, flood control, public water supply, recreation, and more recently environmental protection, should be funded with public funds. This policy with respect to navigation and flood control dates back to the last century. The Reclamation Act of 1902 reflected a policy of developing irrigated agriculture in the 17 western states. More recently our concerns as a society with environmental protection have been backed by massive Federal funding.

These policies of public funding have placed the Congress of the United States through their powers of taxation and appropriation in the position of determining which Federal projects will be built. These projects have been the traditional "pork barrel" of the Congress. In recent years the Carter administration attempted to challenge the traditional "pork barrel" of the Congress by examining various water projects on their merit. This brought on a political outcry from the various parties with vested interests.

To some extent state legislators play a similar role to the U.S. Congress. In California the State has provided massive funding to major water projects. The magnitude of the large interbasin diversion of water is truly enormous. Many of these projects have been funded by state funds; these were only possible in a large and populous state such as California. Again the state legislator plays a key role in determining which project will be funded. In the face of large Federal deficits the Reagan Administration has been seeking to cost share Federal projects, including water projects. The current administration has indicated a willingness to change the priorities for building Federal water projects based upon the willingness of state and local agencies to share the costs. This cost sharing will place state legislators in a more important role as capital investors in water projects.

This aspect of public funding for major projects makes the water industry somewhat different from industries which operate largely with private funds. Public funding inevitably brings with it politics and public participation through the political process. Many technocrats argue that the degree to which water management is a democratic (political) process leads to an inefficient allocation of the resource. To what extent the technocrats are right or wrong is the subject of debate. It seems clear that major policy decisions regarding water in this country reflect our democratic processes at work, as such they are inevitably political—society makes a choice.

Finally concerned citizen interest groups play a considerable role in impacting the decisions of the direct regulators, especially the capital investors. Environmental laws requiring an environmental impact statement for each action which significantly impacts the environment has placed a powerful tool in the hands of concerned citizen groups. Citizen groups can now insist that major actions impacting water resources be openly debated publically before they are taken. This adds to the lead time in taking any action. In some instances it seems to be used purely as a delaying action. In other instances it has led to public referendums concerning particular actions. This is the case in California in which the fate of the Peripheral Canal was decided by a public referendum.

THE SUPPLIERS

One can identify a range of water suppliers, from the individual with his own well, to the Federal Government. It is particularly interesting to look at what motivates the larger suppliers.

The large Federal irrigation projects depend upon creating a user cooperative which contracts to use the water and gradually take over the project. The supplier is usually totally responsive to the needs of the user cooperative. Very often state projects find themselves in the same position. They enter into contracts to supply water, often these are future contracts. Whether the project is built or not depends to some extent upon lining up a group of potential users (Andrews and Sansone, 1983). The user cooperatives play an important role in determining the behavior of the suppliers.

The Bureau of Reclamation was created to build irrigation projects in the arid and semiarid western states. That policy was implemented in 1902 in the Newlands Act which created the Bureau of Reclamation. Certainly the policy has been successful, there are currently approximately 50 million acres under irrigation in the 17 western states. Recently as the available supply of water has become fully utilized, especially in the Southwest, society has begun to question the relevance of a policy of stimulating as well as subsidizing at a national scale, western irrigation. Any questioning of that policy threatens the existence of the bureau of Reclamation.

There has been a recognition by many of the suppliers that the water supply in the West is finite and that having access to an adequate, if not abundant, water supply would be a determining factor to future growth. A number of the urban areas grasped this concept in the early part of this century. Two of the most aggressive urban areas in reaching out for a water supply have been Denver and Los Angeles. They sought to aggressively increase their water supplies not because there was a current shortage, but rather because farsighted urban developers could see that water supply would become a future potential limit to growth. In many cases water supply has been utilized in the West as a lever to consolidate metropolitan government.

In a urban setting sewage disposal is intimately linked to water supply. This link will become increasingly important as reclaimed sewage is used for supply. It is increasingly important to recognize that quality and quantity go hand in hand.

THE USERS

The dominant user of water, especially in the United States, is irrigated agriculture. It has been national policy to stimulate agriculture in the West; we have built numerous irrigation projects with this purpose in mind. The cost of water is a significant factor in whether irrigation is economically feasible. There has been a strong effort on the part of the agricultural community to keep the price of water reasonable, well within the price affordable by agriculture. Since the suppliers and the users are often linked through the user cooperatives, there have been strong incentives by both parties to keep the cost of water cheap. Indeed, our policies as a country of fostering western agriculture have tended to provide subsidies for agriculture, in the process we have maintained a cheap price for water. In a number of instances economists have argued that the price of water to agriculture is well below its true cost. Perhaps, one of the more interesting examples is an analysis of costs and prices associated with the California Water Plan which appears in *Water Supply: Economics, Technology and Policy* by Hirschleifer, et al. (1960).

As water has become fully utilized in the West, especially in the Southwest, the competition for the available supply has increased. Industrial and urban users can afford much higher prices for water than can agriculture. In recent years a number of instances have occurred in which both communities and industries have purchased water rights from agricultural users. In many instances both parties have been happy with the transaction. One of the more interesting instances was a power plant in south-central Utah. The power company purchased water rights from farmers for approximately $1,700 an acre foot, presumably they were willing to go to more than $3,000 an acre toot before abandoning the project.

As competition increases for water, the price, in a traditional economic sense, should increase to the extent that our water institutions and the law allow. Traditional market economics will dictate user behavior. The marketplace may become increasingly important in allocating water.

One has to temper any consideration of the role of the marketplace in water by two other factors:

1. Traditionally we have tended to view water as something special and have intervened as a society by not allowing a market to allocate water.
2. Water quality considerations are not readily accommodated within the market.

These two considerations, especially our concerns with water quality may very well impede market economics from playing a dominant role in our allocation of water.

Economics play an important role in controlling water use in many areas. Increased energy costs along with dropping water levels and deep ground-water pumping have pushed water costs in many areas to the point that irrigation is no longer viable. In Arizona, the U.S. Geological Survey estimates that perhaps as much as 25 to 30 percent of acreage once irrigated is now out of production in many areas. This is simply the consequence of lower prices for certain agricultural commodities coupled with rising water costs.

DISCUSSION OF WATER INDUSTRY MANAGEMENT

In most traditional industries managers are in a position to make decisions regarding both investments and operations. Their concern is: "Can I compete in the market and make a reasonable profit?" The same considerations may exist at the lower levels of the water industry. However, as one proceeds to the larger decisions regarding water, it seems clear that managers do not have similar authority and power to that which exists at similar levels in traditional industries. Water management decisions are dispersed among a number of public agencies (Andrews and Sansone, 1983).

Water management often involves a choice by society. Inevitably in a democracy such as ours, these choices are made as political decisions. The "managers" are usually better described as regulators enforcing the laws established to allocate our water resources.

To a large extent the decision making with respect to water is dispersed. It is interesting to ask to what extent this dispersion is deliberate. One of the more astute state managers I have talked with recently suggested that legislative bodies have deliberately dispersed water authority in order to insure a democratic process. On the other hand, many of the institutions which deal with water have evolved over the past one hundred years; they have evolved in such a way as to distribute the management of water among a number of institutions. One could argue that the dispersion of power is a result of the evolution rather than a conscious decision.

It is difficult to envision the traditional concepts of operations research being applied to problems of water management. There is no one manager with the power to implement an "optimal" solution. One can, however, envision operations research applied to problems of water management in an attempt to explore institutional changes. In addressing problems of water management, a sequence of questions needs to be addressed:

1. Can a particular water-resources system be developed and/or operated in a more efficient manner? Certainly this question can be addressed by many of the techniques of hydrologic analysis coupled with systems analysis and operations research.
2. Assuming the answer to the question above is yes, then a relevant question becomes, do we need an instutional change to make more efficient management possible? Often in water resources an institutional change is necessary.
3. Is an institutional change feasible within the legal and political climate which exists (put another way, are we clever enough to design a feasible institution which will allow the management to be more efficient)?

The question of improved water management becomes one of designing better institutions which are politically and legally feasible.

The use of ground water to supplement the surface water supply in the South Platte Valley of Colorado illustrates how institutions evolve and provide more efficient management. Figure 6 is a plot net economic benefit versus ground-water pumping capacity for a hypothetical reach of the South Platte Valley.

As pumping capacity is installed in the system, the net benefits increase; in fact, the benefits more than double as ground water is utilized (Bredehoeft and Young, 1983). However, in dry periods pumping ground water impacts the streamflow, making less streamflow available for users with senior water rights. The institutions in Colorado now require the ground-water pumpers to collectively augment the streamflow during periods of low flow. They do this by either renting surface-water reservoir storage or through the installation of wells which can be used to pump ground water purely for the purpose of supplementing streamflow. The idea of collective streamflow augmentation by the ground-water users is a relatively new idea. It was only arrived at after

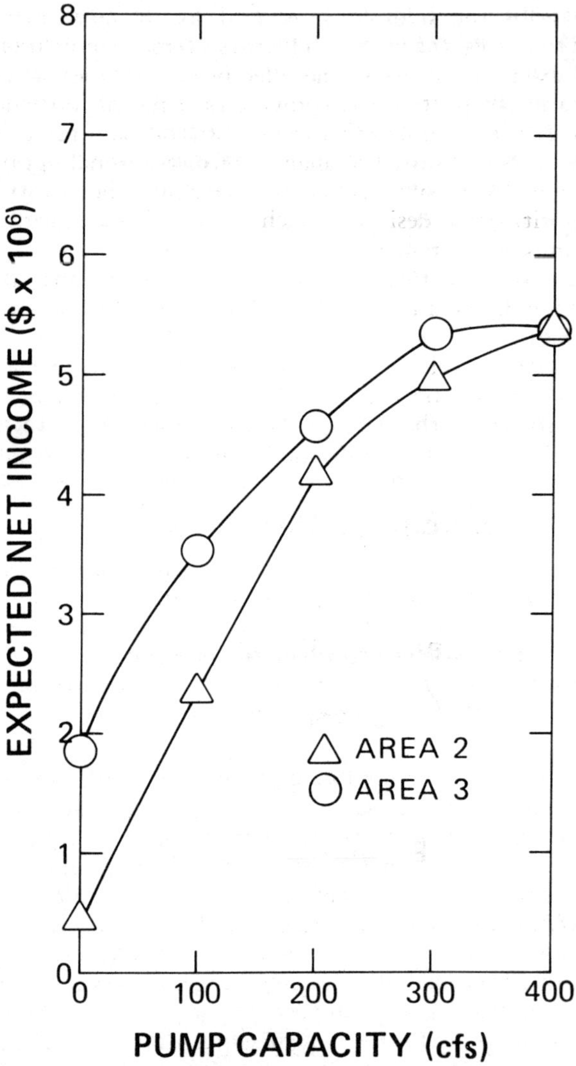

Figure 6. Plot of net economic benefits versus capacity to pump ground water for a hypothetical reach of the South Platte Valley in Colorado.

several unsuccesful attempts by the State Engineer to regulate individual ground-water pumpers.

The conjunctive use management problem along the South Platte system has been the subject of several attempts to simulate the system, both the hydrology as well as the agricultural economics. The analyses of the South Platte system are more useful if viewed as attempts to assess the effectiveness of the current institutions rather than attempts to provide a scheme for "optimal" management. It is my opinion that the analysis of water resources systems is much better suited to analysis of institutional options rather than the design of optimal management schemes. The problem of institutional design is much more difficult since law and politics must almost inevitably be considered.

I would argue that there is a general hierachy of considerations in reaching decision regarding water projects; this hierarchy is shown in Figure 7.

At the lowest level are technical considerations regarding both quantity and quality. Often the technical knowledge is incomplete, partly from the status of both available data as well as the incomplete nature of our current levels of scientific under-

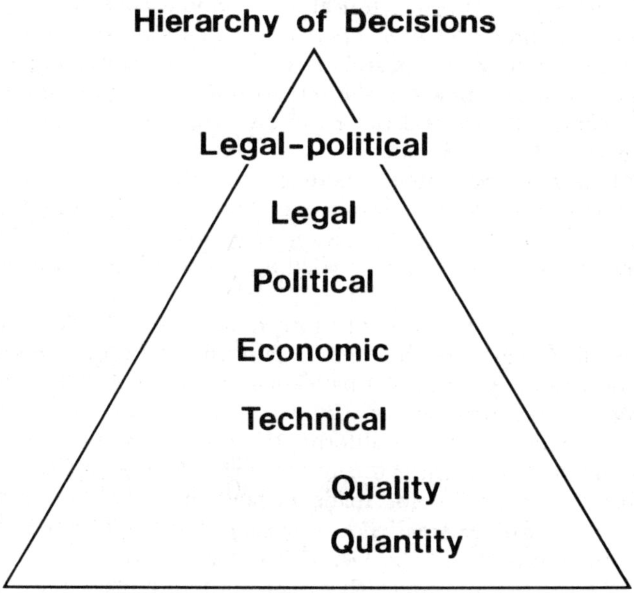

Figure 7. Hierarchy of considerations in reaching decisions regarding water.

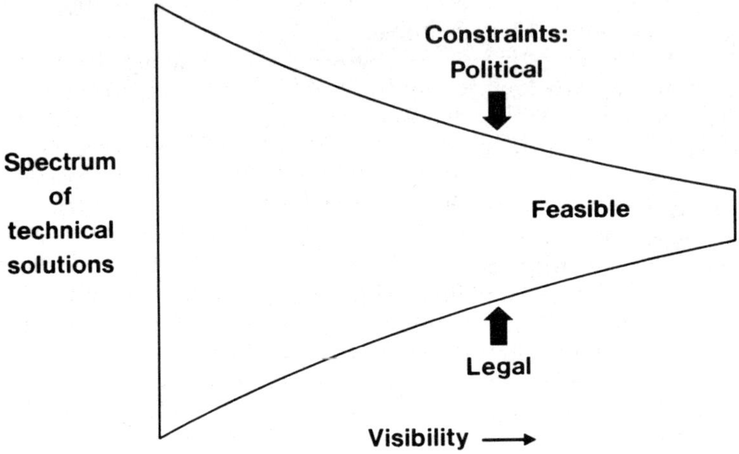

Figure 8. Role of legal and political constraints in restricting feasible solutions as the action increases in visibility.

standing. The technical considerations form a base for evaluating costs and benefits. Both technical and economic considerations fuel the political process. Often a political decision is reached with the constraints dictated by existing law. Finally the law changes, or is changed through new legislation, to reflect new information as well as changing societal values. In essence, this is the democratic process at work in the area of water resources.

One other comment is perhaps worth making. As the scale and cost of projects increases the decisions usually become more political. Commonly the spectrum of feasible solutions is reduced by increased visibility; I have tried to depict this phenomenon on Figure 8.

A recent illustration of this phenomena was the defeat of the Peripheral Canal by referendum in California and the actions which are following. A series of smaller incremental decisions are now being made which will divert water needed to meet contract obligations in Southern California through the Sacramento Delta without the environmental benefits of the Peripheral Canal. Each of these decisions is being made without much public comment. The sum total of the smaller decisions may have much more adverse ecological impacts on the Delta than the Canal.

At the present times there is a shift in power amongst the various "players" within the water industry. Two factors have caused this shift:

1. increased concern with water quality; and

2. the fact that throughout most of the western states the available water is largely allocated and consumed.

The concerns with water quality especially at the Federal level have thrust the Federal EPA into problems of water resources management. State Environmental Agencies, partly through the prodding of the Federal government, now have considerable regulatory authority in the area of water resources management. These are relatively new developments, occurring in the last two decades.

As the available water has become more completely consumed the traditional solutions of building dams and canals to store and redistribute water have diminished in importance. Most students of water resources in this country agree that the era of building large structures to augment the water supply is essentially over in most of the western part of the United States (Englebert 1984). We are now entering an era of increased competition for water. This shift diminishes the importance of the capital investors and the major construction agencies. On the other hand, it increases the importance of the law and the courts since as a society we resort to the courts to resolve questions of competition for water.

ROLE OF HYDROLOGY

What is the message in this analysis of water management for the hydrologist? Certainly one of his primary roles is to understand the physical-chemical-biologic system in sufficient depth so that he can indicate how the system will probably respond to various stresses, either natural or imposed by man. Hydrology is a scientific discipline in its own right; and as such is worthy of scientific investigations. The fact that many of these investigations will have implications for management may place a higher priority for society on these investigations.

As hydrologists our present level of understanding of the broad science of hydrology has many gaps. There is a basic need for much additional research. My own list of scientifically important hydrologic research topics is listed in Table 2.

This is not a small list; these are major gaps in our understanding. Each of these gaps in scientific understanding also has important management implications.

With respect to the issues of water management, my analysis suggests that there is no water "czar" with the power to implement an "optimal" scheme on any but the smaller water resources systems in this country. It seems questionable that worrying about "optimal" management has much relevance in the real world of water resources management in this country. Optimal management may play an important role in the question of exploring institutional changes. However, the real question is one of de-

TABLE 2

Important Areas for Research in the Science of Hydrology

Surface Water

1. Understanding of the transport of:
   a. heavy metals
   b. organic compounds
2. Understanding the link between chemical quality and the aquatic ecosystem.
3. Understanding of extreme events:
   a. floods
   b. droughts

Ground Water

1. Understanding of transport of chemical constituents undergoing chemical reactions.
2. Understanding of the role of microbes in the control of chemical reactions beneath the earth's surface:
   a. unsaturated zone
   b. saturated zone
3. Understanding of transport of chemical constituents through the unsaturated zone.
4. Understanding the hydrology of fractured rocks:
   a. flow
   b. transport

veloping institutions which move water resources management toward more "efficient" solutions.

Water management has evolved in this country as a largely democratic process. That choice seems to me a deliberate one on the part of our society. This process has allowed rather full utilization of our water resources, especially in the west. It has led to a high degree of management. The role of hydrology in the issues of management in this country, is to enable society to make as informed a choice as is possible.

References Cited

Andrews, B. T. and M. Sansone (1983). Who runs the rivers? Dams and decisions in the new west. Stanford, CA: Environmental Law Society, Stanford Law School, 452 pp.

Bain, Joe S., Richard E. Caves, and Julius Margolis. *Northern California's Water Industry.* Baltimore: The Johns Hopkins Press, 1966, 766 pp.

Bredehoeft, J. D. (1984). Physical limitations of water resources in *Water Scarcity.* A. E. Englebert (editor). Berkeley, CA: University of California Press, 17-44.

Bredehoeft, J. D. and R. A. Young (1983). Conjunctive use of ground water and surface water for irrigated agriculture: risk aversion. Water Resources Research, v. 19, 1111-1121.

Engelbert, A. E., editor (1984). *Water Scarcity.* Berkeley, CA: University of California Press, 484 pp.

Hirschleifer, J., J. C. DeHaven, and J. W. Milliman (1960). *Water Supply: Economics, Technology and Policy.* Chicago: University of Chicago, 378 pp.

Lucky, R.R., E. D. Gutentag, and J. B. Weeks (1981). Water level and saturated—thickness changes, predevelopment to 1980, in the High Plains Aquifer in part of Colorado, Kansas, Nebraska, New Mexico, Oklahoma, South Dakota, Texas, and Wyoming. U.S. Geological Survey Hydro. Invest. Atlas 65V.

National Research Council (1977). Energy and Climate: Studies in Geophysics. Washington, D.C.: National Academy of Sciences, 158 pp.

National Research Council (1984). Groundwater Contamination: Studies in Geophysics. Washington, D.C.: National Academy of Sciences, 179 pp.

Pye, V. I. and J. Kelley (1984). The extent of groundwater contamination in the United States. In Groundwater Contamination. Washington, D.C.: National Academy of Sciences, 23-33.

State of California (1979). The California water atlas: State of California. Sacramento, CA: Governor's Office of Planning and Research and Dept. of Water Resources, 118 pp.

U.S. Water Resources Council (1978). The nation's water resources, 1975-2000. Second National Water Assessment, V.1: Summary, 86 pp.

JOHN D. BREDEHOEFT
U.S. Geological Survey, Menlo Park, CA 94025

Dr. John D. Bredehoeft has been a research geologist with Water Resources Division of the U.S. Geological Survey for more than 20 years. He has worked on a variety of groundwater problems, including numerical models for the analysis of both flow and transport in groundwater systems, analytical methods for the field determination of aquifer parameters, as well geophysical experiments for both the prediction and control of earthquakes. He spent two years at Resources For The Future, on loan from the Geological Survey, where he was involved in analytical studies of the economics of groundwater management. He has received both the Meritorius and Distinguished Service Awards of the Department of the Interior; the Horton Award of the American Geophysical Union; the Meinzer Award of the Geological Society of America; and the Boggess Award of the American Water Resources Association. He served as President of the Hydrogeology Division, Geological Society of America; has been a member of numerous committees and panels of the National Academy of Sciences; and is a Fellow of the American Geophysical Union.

For five years Dr. Bredehoeft managed the entire water research activities of the U.S. Geological Survey; he also served as Regional Hydrologist, Western Region, a position in which he was responsible for all of the Survey's water activities in the eight western states. Currently, he is a senior research scientist, in Menlo Park where he is actively involved in experiments utilizing water wells as strain-meters in an effort to predict earthquakes.

Dr. Bredehoeft received his BS degree in Geological Engineering from Princeton, and MS and PhD degrees from the University of Illinois. He taught one year at the University of Illinois, as a visiting professor; he currently teaches as a visiting scholar at Stanford.

FOREWORD

A definition of *hydrology* states that it is the science which studies the interaction on planet Earth between hydrosphere on the one hand, and lithosphere and biosphere on the other. Of course, atmosphere connects the other three "spheres" that participate in these complex natural phenomena. Having neglected man in the definition of hydrology, we submit that *water resources* is the study of man's intervention in the hydrological cycle.

Since the Age of Enlightenment—and perhaps even earlier—scientific efforts, i.e., understanding nature, were focused on increasingly sharper definitions of the issues studied. This approach yielded spectacular scientific, technological, and social advances, yet contributed substantially to the present compartmentalization of science. As a result, each part, or even every sub-component, of a phenomenon is treated separately, thus obscuring the interactions between the components. Hydrology was not immune to this process, and we often refer to surface hydrology as if streams and rivers flow in total isolation from ground waters.

We have reached the point where in order to understand hydrology better and to produce more effective water resources systems, we need to study the interactions between the components of the hydrological cycle. To do so, we build models—conceptual constructs—at various scales which, in order to enhance our understanding of nature, need to be integrated within ever-broadening limits. Eventually, global-scale hydrologic models will emerge that will help explain and forecast, for example, streamflow variability in the Eastern Mediterranean as related to precipitation patterns in Sudan. An important detail is the non-linearity often observed when hydrosphere, lithosphere, biosphere and atmosphere interact.

There is at least one important practical outcome of well-constructed macro-scale hydrological models. The biosphere, and in particular the flora, is a huge pumping mechanism transferring very large amounts of water from the land to the atmosphere through transpiration. A model supported by adequate field data can yield reliable estimates of the effects of reducing the forested area in central India, for example, on the discharge and water quality in rivers arising in the Deccan plateau. These estimates would be of fundamental importance to the planning of the development of some major river basins in India.

The Fourth Chester C. Kisiel Memorial Lecture addresses these issues, emphasizing the need and usefulness of global-scale numerical models of the interactive physical, biological and chemical systems on earth. The development and use of these models in connection with good data will ensure that hydrology will be increasingly relevant to man and society.

Nathan Buras

REPRINTED FROM WATER RESOURCES RESEARCH 22 (9), 6S-14S, 1986

THE EMERGENCE OF GLOBAL-SCALE HYDROLOGY

Peter S. Eagleson

GLOBAL ENVIRONMENTAL CHANGE

The atmosphere, hydrosphere and surface layers of the Earth have arrived at their present characteristics through a coevolution of living and non-living components. The picture as revealed by paleoclimatologists is one of large scale natural processes undergoing cycles of dynamic change on a wide spectrum of time scales, from years to hundreds of thousands of years, accompanied synergistically by the evolutionary development of life forms. An example of the evidence for this natural change is offered by the fossil pollen record in North America since the peak of the last ice age 18,000 years ago as determined by Webb et al. This is presented in Figure 1 as taken from Kerr (1984). An increase in summer solar radiation and the retreat of the ice sheet caused the oak and northern pine forests to withdraw to the north and at the same time developed our southern pine forests.

Humans have been altering the environment over large geographic areas for over ten thousand years through their domestication first of fire and then of plants and animals (Sagan et al. 1979). Early civilizations destroyed the temperate forests of China and the Mediterranean Basin, and modern civilizations have greatly reduced the temperate forests of Europe and North America.

In the last five hundred years, however, the hand of man has been increasingly felt on the biogeochemical cycles that control the Earth's metabolism. Energy production, farming, urbanization and technology have altered the albedo of Earth, the composition of its soil and water, the chemistry of its air, the amount of its forest, and the structure and diversity of the global ecosystem. More than thirty percent of the Earth's land area is now under the active management of man with more than ten percent being under cultivation. Chemical compounds having no analogs in nature are being introduced into both air and water at increasing rates.

Most recently the tropical forests have come under attack. Meyers (1979) estimated that Latin America has lost thirty-seven percent of its original rain forest (largely to agricultural development), Southeast Asia has lost thirty-eight percent (principally to logging), and Africa has lost over fifty percent (primarily to slash-and-burn agriculture).

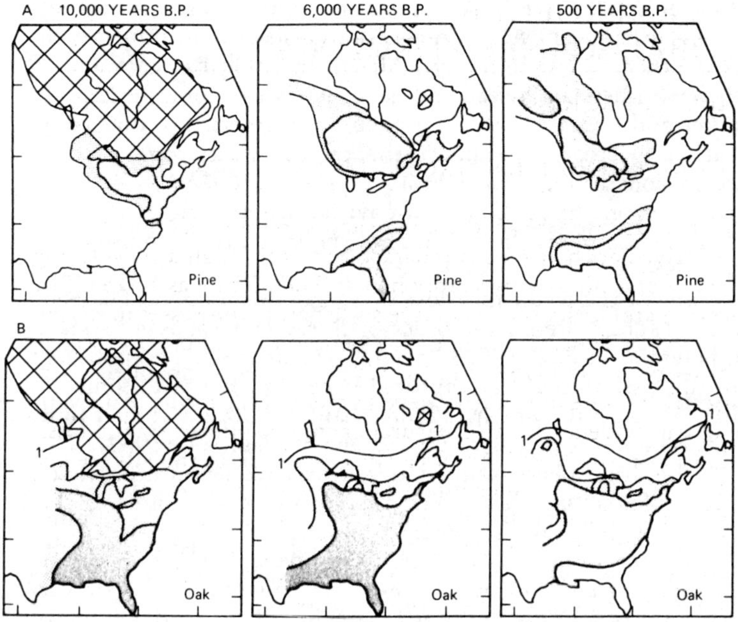

Figure 1. Vegetation Change in Eastern North America During the Past 10,000 Years {from Kerr, 1984} (Crosshatched area is ice sheet. Contour is 1% oak. Light stippling is 5 to 20% oak or 20 to 40% pine. Heavy stippling is greater than 20% oak or greater than 40% pine. Reproduced with the permission of the American Association for the Advancement of Science.)

The alteration of ground cover affects surface albedo and runoff, changes the ratio of sensible to latent heat transport, alters surface winds and erosion rates, and changes the thermal and moisture state of the surface. The microclimates of forested and cleared areas differ markedly. In tropical regions such as the Amazon basin where soils are typically poor, their exposure to sunlight may produce chemical and structural changes that inhibit either agriculture or reforestation and introduce erosion due to the heavy precipitation. In subtropical regions, such as central Africa, where precipitation is limited, a forest ecosystem appears to be unstable (Eagleson and Segarra, 1985) and its destruction leads to a stable tree/grass savanna. Such has been the fate of forty percent of the African equatorial forests as a result of slash-and-burn agriculture (Phillips, 1974).

The global cycle of water is perhaps the most basic of all the biogeochemical cycles. In addition to its strong influence on all the other cycles, it directly affects the global circulation of both atmosphere and ocean and hence is instrumental in shaping weather and climate. Planning and/or construction is underway on various macroengineering weather projects which through their modifications of regional hydrology promise to contribute their own distortions to the course of environmental change.

One example is the drainage of the immense swamps of the White Nile's Sudd region in order to capture for downstream uses some of the water now lost by evapotranspiration. The permanent swamps are on the order of thirty-four thousand square kilometers in surface area and if solely the dry-season evaporation from this surface could be captured it would amount to some 25 x 10^9 m^3 annually which is more than the current annual flow of the White Nile at Khartoum (Chan and Eagleson, 1981). The loss of this atmospheric water and its associated latent heat would surely be felt climatically. The first phase of this project, the 360 km Jonglei canal is nearing completion.

Another project is the diversion of several Soviet rivers away from their current northward flow. The project has two parts, a European portion now under way which will divert the Sikhona and Onega Rivers southward, away from the White Sea, to irrigate 2.5 million acres in the northern Caucasus, and a Siberian portion which if undertaken would send the Ob and Irtysh rivers to the arid regions around the Aral Sea instead of into the Kara Sea. By depriving the Kara Sea of a large fresh water inflow, this latter diversion has the potential for reducing ice cover and thus decreasing the regional albedo.

Both the deforestation and the proposed macroengineering projects act to create anomalous regional moisture and/or heat sources (or sinks) the effects of which may, in theory at least, propagate to distant regions via atmospheric dynamics (Webster, 1982). As possible examples of such "teleconnections," as they have come to be called, we cite first the striking negative correlation between the winter snow cover over Eurasia and the intensity of the following summer monsoon in India. As pointed out by Walsh (1984) this inverse relation (see Fig. 2) is consistent with the argument that widespread snow cover leads to lower springtime air temperatures and hence to higher sea level pressures over southern Asia which oppose the normal monsoonal pressure gradients. Of course, the correlation does not establish causality. Indeed, in atmospheric general circulation model (GCM) experiments Shulkla (1975) related the decrease in Indian monsoon

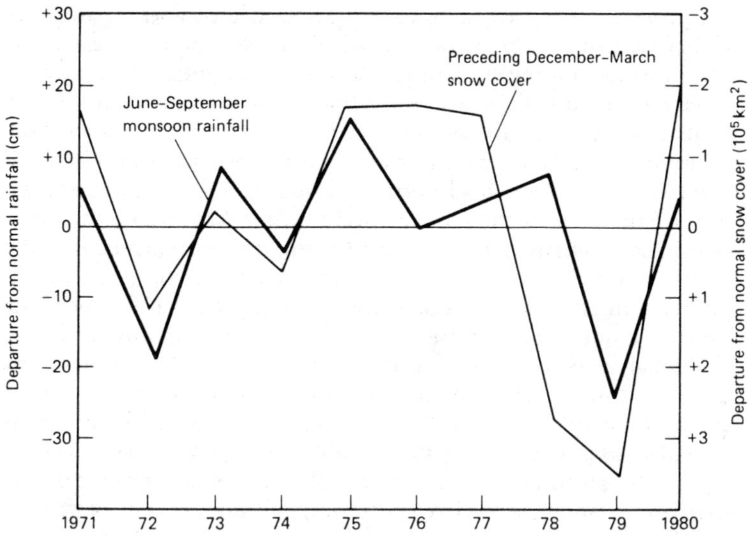

Figure 2. Indian Summer Monsoon Rainfall and Himalayan Snow Cover of Preceding Winter (from Walsh, 1984) (Reproduced with the permission of American Scientist.*)*

rainfall to cold surface temperatures in the Arabian Sea. A similar correlation has been found both observationally and with GCM experiments between drought in northeast Brazil and positive sea surface temperature anomalies in the tropical Atlantic.

Much similar evidence has been assembled to support teleconnections between sea surface temperature anomalies in the eastern tropical Pacific Ocean (El Niño) and middle-latitude atmospheric circulation in the winter hemisphere (see for example Horel and Wallace, 1981); and between sea surface temperature anomalies in the Atlantic Ocean off West Africa and sub-Saharan drought (Lamb, 1983).

The evidence is overwhelming that regional anomalies in the surface state of the Earth as given by its albedo, temperature and wetness have local and sometimes also far-reaching effects upon the atmospheric temperature, humidity and precipitation.

But man's effect on the hydrologic cycle is not limited to these physical issues. His use of the atmosphere for disposal of civilization's gaseous wastes has altered the chemistry of precipitation with serious consequences for fish and other aquatic organisms, crops, forests, wetlands, soils and even buildings. There is a potential here for damage to human health as well and this is beginning to attract serious study (Maugh, 1984). The acidification of water supplies brings increased concentration of potentially toxic metals such as lead, cadmium, mercury and aluminum in that water; the metals are leached from the soil and from sediments and from the pipes and fixtures used in water supply systems. Of particular concern are lead, which is in widespread use as a liner in the cisterns of rural roof catchment systems, and aluminum, which comprises about five percent of the Earth's crust.

Aluminum is practically insoluble in water of neutral or high (alkaline) pH and thus has not been historically available biologically. Within the last decade however high concentrations of aluminum have been found in brain, muscle and bone tissues of patients suffering from dementia and bone disorders and who have also been under long-term dialysis at centers where there is significant aluminum in the water. With the advent of nuclear magnetic resonance (NMR) scanning, high concentrations of aluminum have been found in the brain tissue of many patients with Alzheimer's disease and with senile dementia. Autopsies on victims of certain other central nervous system disorders at isolated locations having abnormal incidence rates have shown similar high concentrations of aluminum. Whether there proves to be a causal relation in these examples or not, the specter of unsafe drinking water adds further motivation to understand the pathways for the global dispersal of atmospheric pollutants.

QUESTIONS OF LARGE-SCALE HYDROLOGY

The case for global-scale hydrology can be made at small scale. Consider the question of the local environmental impact from local land surface change. Will drainage of the swamps reduce the local precipitation? The strength of the local coupling between evapotranspiration and precipitation was estimated at about 10 percent by Budyko and Drozdov (1953) for the European U.S.S.R., and at about the same percentage by Benton et al. (1950) for the Mississippi valley. However, Letau et al. (1979) found places in the Amazon basin where as much as 71 percent of the precipitation appeared to come from locally evaporated water. Salati and Vose (1984) estimate 48 percent recycling for the Amazon Basin as a whole. As concluded by Shukla and Mintz (1982), evaporation change can affect local precipitation but the strength of the recycling will vary from region-to-region depending on how the large-scale circulation is modified. This can be determined only through tracer experiments or by using global-scale modeling.

An allied question seeks the geographical influence function of a local land surface change. That is: What locations will feel the effects of a land surface change here? Reduction of evaporation in the Sudd will reduce the precipitation where? By how much? This calls for tracer studies in global-scale models.

The inverse of this question is of interest for those concerned with identifying the source of their precipitation. That is: Where was the water last evaporated that falls locally as precipitation? Again we need global-scale models to define this atmospheric moisture replacement distance. As was pointed out by Eagleson (1982), the lateral scale of a proposed land surface change will have to exceed this replacement distance before the feedback loop can close to create a downwind amplification of the original disturbance.

These hydrologic scales and feedbacks are seasonally as well as geographically variable. During the winter months, the continental land surfaces are net sinks for atmospheric moisture picked up over the oceans, while in the summer, when thermal convection is the primary precipitation mechanism, the depletion of soil moisture by evaporation and transpiration transforms the continents into net sources of atmospheric water. Understanding these scales is critical to forecasting the location, size and strength of anomalies in the cycle and in defining the environmental impacts of land surface changes. These scales are largely unknown and should be determined for all regions of the globe and for all seasons of the year.

The hope for significant improvement in the accuracy and lead time of local long-range hydrologic forecasting, so important to agriculture, lies in establishing teleconnections to the climatic flywheel—the oceans. The GCM with coupled dynamic ocean provides these teleconnections implicitly.

Conditioned as we are by the traditional engineering demands of water supply and flood protection, hydrologists often lose sight of the broad definition of their field (Federal Council of Science and Technology, 1962) which includes that part of the hydrologic cycle involving the oceans. Actually, the distribution of precipitation and evaporation over the ocean plays an important role in establishing ocean circulation and hence global climate. An example of this is the formation of deep water in the northern Atlantic Ocean.

The best estimates available suggest that evaporation exceeds precipitation and continental runoff on the North Atlantic Ocean and its adjacent seas by about 15%, this deficit being replaced by ocean circulation. The excess evaporation results in a salinity and hence density increase which must be balanced by an exchange for less salty water from another ocean. It is thought that this happens by a sinking and southward flow of the saline surface waters in the North Atlantic accompanied by a shallow northward return flow of less salty water from the Antarctic. Warmed as it passes through the tropics, this returning surface water carries heat to the North Atlantic and upon evaporation transfers much of this to the atmosphere where it becomes responsible for the moderate climate of northern Europe.

Our quantitative knowledge of the oceanic branch of the global hydrologic cycle is quite poor. What are the water balances of the various ocean basins? We have very poor knowledge of the oceanic fluxes due to precipitation and evaporation let alone those due to continental groundwater discharge and to sea floor vents. Observational difficulties suggest that in the short term at least global models will provide our best estimates of oceanic precipitation and evaporation.

GLOBAL HYDROLOGIC MODELING

Atmospheric general circulation models are based on the fundamental equations that describe the dynamics and energetics of fluid motion. These include the equations of motion (conservation of momentum), the first law of thermodynamics (conservation of energy), the continuity equations for air mass and water vapor (conservation of mass), and the ideal gas law (approximate equation of state). These equations are solved numerically on a grid having a horizontal resolution on the order of 5° (i.e., several hundred km) and with as many as 12 vertical layers up to an atmospheric limit of say 10 mb. The computational time step at this resolution is about 7 minutes. Of course each of the "prognostic" (i.e., independent) atmospheric variables: wind, temperature, pressure or density, and humidity must be given an initial condition at each node of the solution net, and a boundary condition at each surface node. Early models prescribed fixed boundary values but more recently interactive boundary conditions of progressive sophistication have been introduced; first for the land surface (about 15 years ago), second for the oceans (about 5 years ago), and currently for the vegetation. For example, at the ground surface current GCMs calculate the temperature and soil moisture concentration using approximations of the surface heat and water balances. Other "diagnostic" variables such as cloudiness, precipitation, surface radiative flux, surface sensible heat flux, evaporation, runoff and snow cover are estimated from parameterizations

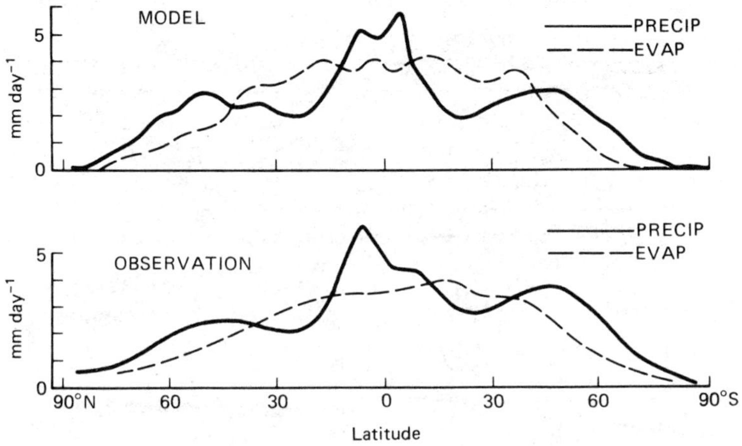

Figure 3. Zonally-Averaged Mean Annual Precipitation; Comparison of Model and Observations (from Mitchell, 1983).

which relate these subgrid-scale processes to the large-scale variables that are resolved by the model (Gates, 1983).

After the global solution of these equations has been advanced in time for a period of perhaps several years, the time-averaged prognostic variables approach constant values which define the model climate. At this time comparisons with global distributions of observed average annual and average seasonal quantities can be made. At the current state of model development typical comparisons of zonally-averaged (i.e., circumferentially-averaged) annual average precipitation and evaporation agree quite favorably with observation (Mitchell, 1983, see Fig. 3) and the global distribution of the local annual averages shows basic agreement in the location; if not in the intensity, of regions of high and low precipitation (Hansen et al., 1983, see Fig. 4). There are local discrepancies of up to 100% in annual totals mostly in tropical regions. The models are also used to simulate the average seasonal cycle of precipitation including monsoons and the movement of the tropical rain belt following the ITCZ.

The atmospheric GCMs are capable of realistically simulating the interannual variability of the hydrologic cycle also because of the unstable transient cyclones arising primarily at mid-latitudes in solution of the equations of motion.

There is much room for improvement in the formulation of GCMs, particularly in the parameterization of subgrid-scale hy-

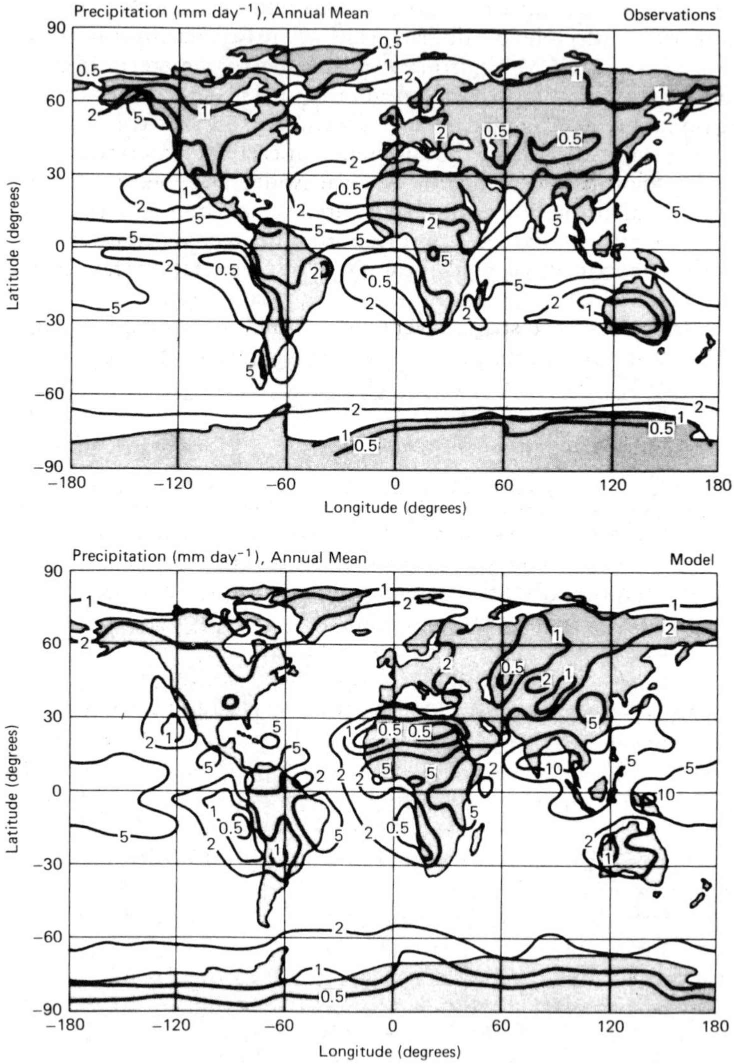

Figure 4. Global Distribution of Mean Annual Precipitation; Comparison of Model and Obvservations {from Hansen et al., 1983}. (Reproduced with the permission of the Monthly Weather Review.*)*

drologic processes but there is also need for additional basic understanding of some critical hydrologic phenomena. For example:

Precipitation arising from moist convection is acknowledged to be spatially variable at subgrid-scale when calculating the mass falling on the GCM gridsquare. The usual parameterization of land surface hydrology then smears this mass uniformly over the gridsquare when calculating the subsequent soil moisture fluxes. This simplification made no difference in early parameterizations of the Thornthwaite-Budyko (Thornthwaite and Mater, 1955) type in which the flux is linearly related to soil moisture concentration. Current parameterization efforts are directed toward incorporating more realistic non-linear moisture flux relations however, whereupon the spatial averaging question becomes crucial. How do we represent the spatial average dynamic hydrologic behavior of mesoscale areas in the presence of inputs and physical parameters which are spatially variable at smaller scale and in a manner which is at best only generically known? This is an unsolved problem that arises wherever in nonlinear dynamics disparate scales must be coupled.

Vegetation cover has a profound influence on the heat and moisture budgets of the land surface and yet in current GCMs it is a prescribed boundary condition. Such prescription does not account for the synergism among climate, soil, and vegetation that determines such parameters as canopy density and type and hence albedo and water use. Of particular importance in this regard is the prognostic distinction between deciduous and evergreen vegetation. It is thus important to develop and use in GCMs vegetation models which are truly interactive. This is beginning insofar as the water use and albedo of prescribed vegetation types are concerned. If the interaction is to include model specification of vegetation type, however, it will first be necessary to understand the climate and soil conditions that determine one type in preference to the others. There is ample empirical evidence (e.g., Perrier, 1982, see Figure 5) that the primary types of world vegetation are arranged, to a significant degree, according to variations in the availability of water and energy, contemporary understanding of the role of humans, pests, and fire notwithstanding. Ultimately the soil too should be made interactive as its physical and chemical character are part of the synergism. As modeling of global biogeochemical cycles moves into the planning stages this becomes a serious consideration (NRC, 1985).

Location of the time-varying snow line and sea ice boundary is critical to the magnitude and distribution of Earth's albedo. Additional work is needed on the snow and ice melt problems to be able to define these boundaries with reasonable accuracy.

Coming back now to the spatial scale of the hydrologic cycle

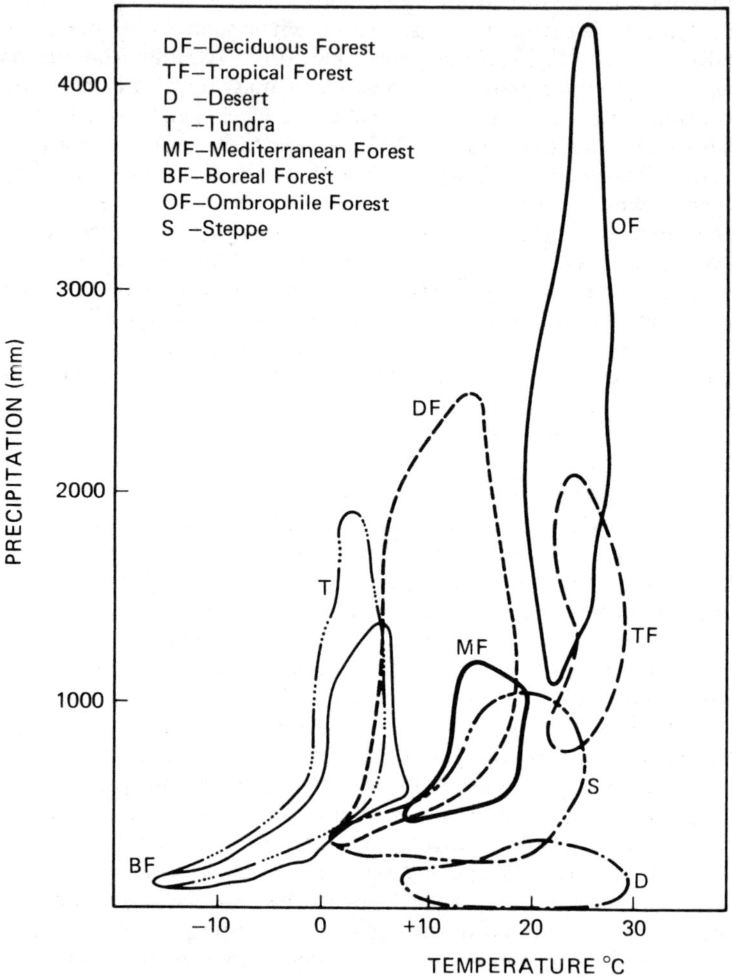

Figure 5. Climatic Limits of Major Zonobiomes (from Shuttleworth, 1982). (Reproduced with the permission of Cambridge University Press.)

and its exploration using atmospheric GCMs, we will present some preliminary results of Koster (1985) as a demonstration of the power and utility of these models as a hydrologic research tool. Using the GCM of the NASA Goddard Institute of Space Studies (GISS) with medium resolution (8° x 10° grid), Koster (1985) "tagged" the water in a one-day impulse of evaporation

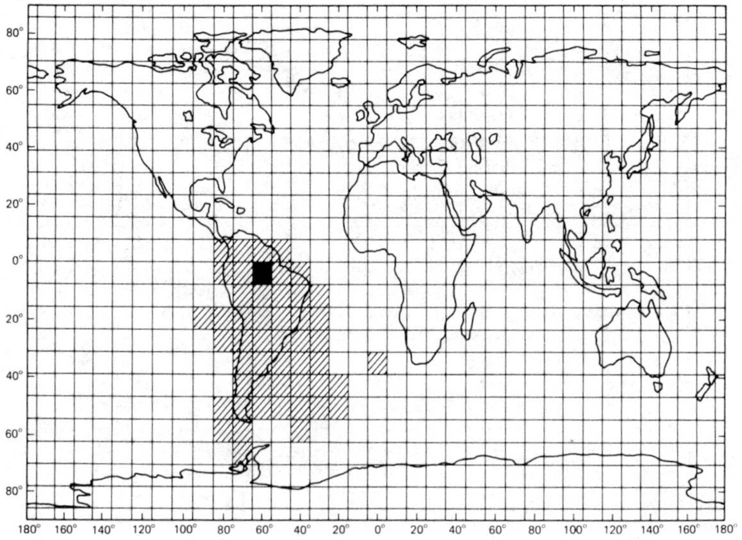

Figure 6. Region of Influence of Amazon Evaporation {Koster, 1985}.

from selected grid squares and followed this water for two months to see where it precipitated. The GCM initial conditions were those corresponding to a particular month. The characteristic time for precipitation of the evaporated water varied from 2 to 5 days for the grid squares tested. Only three of the most environmentally interesting grid squares are presented here.

Figure 6 follows the water evaporated in March from the grid square most closely representing the Amazon basin and marked here by the solid shading. The lighter shaded grid squares show where most of the evaporated water was subsequently precipitated. Grid squares receiving less than an arbitrary small amount are considered to have received none. Notice in this case that the evaporation apparently gets caught in the Southern Hemisphere Hadley cell and is carried primarily to the south with relatively little east-west dispersion. This preliminary study indicates that the South American continent would be the primary area affected by precipitation change due to Amazon deforestation. The study also shows that 37% of the water evaporated from the Amazon basin in March is recycled as subsequent precipitation on the same grid square.

Figure 7 follows in the same fashion the water evaporated in March from another site of extensive deforestation, Southeast

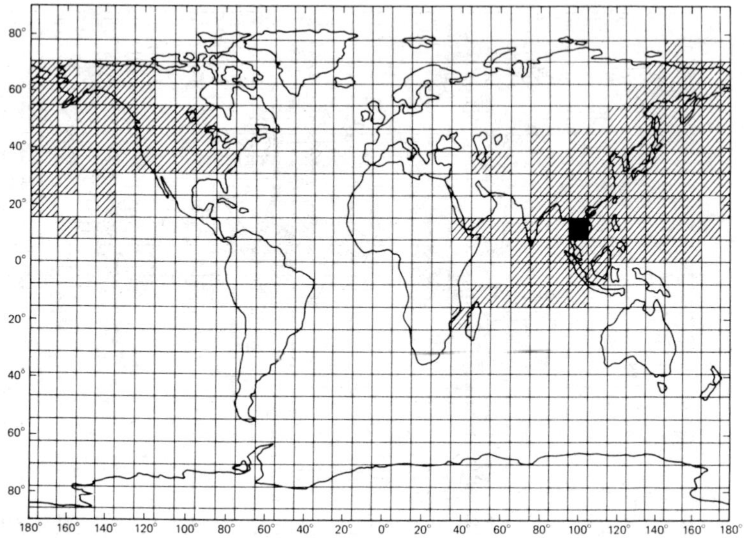

Figure 7. Region of Influence of S.E. Asia Evaporation {Koster, 1985}.

Asia. In this case there is a strong west-to-east advection of the moisture added to the expected poleward movement. The "influence radius" of this location is enormous as far as evaporation is concerned, and 52% of the evapotranspiration is recycled into local precipitation.

The final evaporation example is that of Sudan's Sudd region discussed earlier and is presented here as Figure 8. The precipitation resulting from January evaporation is largely confined to the African continent with some being advected onto the Atlantic Ocean by the Easterly winds of these latitudes. About 19% of the Sudd evaporation during this month falls back on the Sudd as precipitation.

These results are far from definitive of course; being impulsive rather than steady state; being for only one season of the year; being for only a single sample of the possible initial conditions; and most importantly, being subject to all the approximations and inaccuracies of GCMs at their current state of development. At the least, however, they do have qualitative, comparative values; they are eloquent testimony to the potential utility of these models in hydrology; and they serve as valuable guides to the design of field programs.

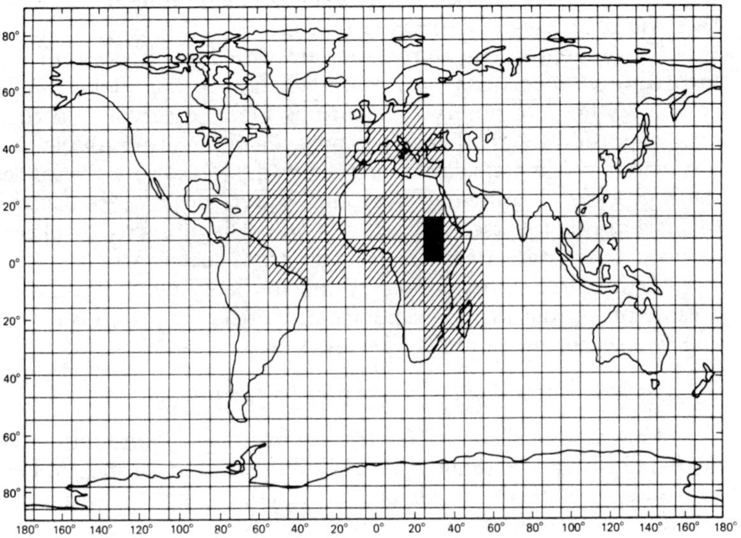

Figure 8.. Region of Influence of Sudd Evaporation {Koster, 1985}.

EXPERIMENTAL SUPPORT

Science advances on two legs, analysis and experimentation, and at any moment one is ahead of the other. At the present time advances in hydrology appear to be data limited; not the micro-measurements of the laboratory taken to learn about the one-dimensional physics of isolated processes, but rather those macro-scale field observations needed to understand the hydrologic coupling of mesoscale precipitation events with heterogeneous land surfaces, and the global-scale assessments necessary for monitoring the changing inventory of Earth's waters.

We have already mentioned the importance of the mesoscale measurements to subgrid-scale parameterization in GCMs and of course they lie at the core of conventional catchment hydrology as well. Measurements of the bulk energy and water fluxes over inhomogeneous mesoscale areas (i.e., 10^2-10^4 km^2) are badly needed to learn how best to parameterize these fluxes in the presence of different vegetation types and in different climates. Planning is underway for such experiemnts by two international groups (IAMAP and COSPAR, 1983; WMO/ICSU, 1984) and hopefully the U.S. interagency STORM project (1984) will be modified to include these hydrologic objectives. It is important that hydrologists play an active role in both the planning and the conduct of these enormously expensive experiments to ensure that the broadest objectives are met.

The inventory measurements are called for because of our surprisingly poor quantitative knowledge of the global water cycle. We need to observe the various global reservoirs on time scales appropriate to their dynamics: days for atmospheric and soil water; weeks for lakes and snow pack; weeks to years for sea ice; and years to millenia for ground water and glaciers. Clearly this is a formidable task of great cost and calls for increased efforts to make these measurements possible from space.

A third area where observations are badly needed is that of oceanic precipitation and evaporation. Here again progress is limited by lack of measurement technology.

EDUCATIONAL IMPLICATIONS

The development of GCMs has quite appropriately been carried out by meteorologists, climatologists, and oceanographers who first foresaw the need for and potential of the global-scale approach to study climate change and to improve long range weather forecasting. Their early assumption that the land surface was a passive and weak participant in the atmospheric action led to a hydrologic parameterization consisting of prescribed surface moisture state, either bone dry or saturated, and produced an over-active model hydrologic cycle. Subsequent numerical experiments demonstrated the high sensitivity of model climate to the land surface moisture state and brought concerted effort to incorporate more realistic hydrologic algorithms within the very real (but continuously expanding) constraints of computation time. This effort has come primarily from within the meteorological community and from physicists interested in achieving the potential for remote sensing in this application. Sincere attempts to involve hydrologists have been largely unsuccessful.

Hydrologists have much to offer this modeling effort and have even more to gain by being active participants. They bring an accumulated experience with the hydrologic behavior of inhomogeneous, mesoscale catchments that allows them to define the most important parameters and processes in specific climatic and geologic circumstances. More importantly perhaps they bring the engineering motivation for solving the problems of people and an understanding of the water needs of man's agricultural, urban and industrial life support systems. Hydrologists should know the important environmental questions to be asked of a verified model, and their patticipation in model development will help ensure the model's ultimate capability to be of appropriate benefit.

To be effective in such an interdisciplinary partnership the hydrologist will need a familiarity with subject areas that are seldom a part of his current educational program. These include radiation physics, planetary fluid dynamics, precipitation processes, micrometeorology, plant physiology, natural and managed ecosystems, and the analysis of random fields—a challenge that would surely have appealed to Chester Kisiel.

SUMMARY AND CONCLUSION

Because of humanity's sheer numbers and its increasing capacity to affect large regions, the hydrologic cycle is being altered on a global scale with consequences for the human life support systems that are often counterintuitive. There is a growing need to assess comprehensively our agricultural, urban and industrial activities, and to generate a body of knowledge on which to base plans for the future. It seems safe to say that these actions must come ultimately from global-scale numerical models of the interactive physical, chemical and biological systems of the earth. Of central importance among these systems is the global hydrologic cycle and its representation in these models presents many analytical and observational challenges for hydrologists.

We must devote more attention not only to the technical issues of hydrology raised by the model builders but also to encouraging and preparing more young hydrologists to build a career in this direction. He who controls the future of global-scale models controls the direction of hydrology.

ACKNOWLEDGMENT

The author wishes to thank his colleagues on the U.S. Committee for an International Geosphere-Biosphere Program whose thoughts and even words may appear in this paper without specific attribution.

References Cited

Benton, G. S., R. T. Blackburn and V. O. Snead, "The role of the atmosphere in the hydrologic cycle, *Trans AGU,* 31(1), 61-73, 1950.

Budyko, M. I. and O. A. Drozdov, Characteristics of the moisture circulation in the atmosphere, *Izv. Akad. Nauk SSSR Ser. Geogr. Geofiz.,* 4, 5-14, 1953.

Chan, S-O. and P. S. Eagleson, Water balance studies of the Bahr el Ghazal Swamps, *MIT Department of Civil Engineering, R. M. Parsons Laboratory Report* No. 261, 1981.

Dey, B. and O. S. R. U. Branu Kumar, Himalayan winter snow cover area and summer monsoon rainfall over India, *J Geophys. Res.,* 88, 5471-5474, 1982.

Eagleson, P. S., "Hydrology and Climate," in *Scientific Basis of Water Resource Management,* 127 pp., Geophysics Study Committee, Studies in Geophysics, National Academy Press, Washington, D.C., 1982.

Eagleson, P. S. and R. I. Segarra, Water-limited equilibrium of savanna vegetation systems, *Water Resources Research* (in press), 1985.

Federal Council of Science and Technology, Committee for Scientific Hydrology, *Trans. AGU,* 43(4), p. 493, 1962.

Gates, W. L., The use of general circulation models in the analysis of the ecosystem impacts of climatic change, *Report of the Study Conference on Sensitivity of Ecosystems and Society to Climatic Change,* World Climate Impact Studies Program, WMO, Geneva, 1983.

Hansen, J., G. Russell, D. Rind, P. Stone, A. Lacis, S. Lebedeff, R. Ruedy and L. Travis, "Efficient three-dimensional global models for climate studies: models I and II, *Monthly Weather Review,* 111, pp. 609-662, 1983.

Horel, J. D. and J. M. Wallace, Planetary scale atmospheric phenomena associated with the interannual variability of sea-surface temperature in the equatorial Pacific, *Mon. Wea. Rev.,* 109, pp. 813-829, 1981.

Interagency Team for STORM Central, "The National STORM program STORM-central phase, preliminary program design, NCAR, Boulder, 147 pp., 1984.

International Association of Meteorology and Atmospheric Physics (IAMAP), and Committee on Space Research (COSPAR), Development of the implementation plan for ISLSCP, phase 1," Report to the UN Environmental Program (UNEP) Project No. FP/1303-83-03 (2426), Paris, 1983.

Kerr, R. A., Climate since the ice began to melt, *Science,* 226, 326-327, 1984.

Koster, R. D., The global hydrologic cycle: local recycling, characteristic times, and horizontal scales," unpublished Ph.D. thesis, MIT Department of Civil Engineering (in preparation), 1985.

Lamb, P. J., Large-scale Tropical Atlantic surface circulation patterns associated with Subsaharan weather anomalies, *Tellus,* 30, 240-251, 1978.

Lettau, H., K. Lettau, and L. C. B. Molion, Amazonia's hydrologic cycle and the role of atmospheric recycling in assessing deforestation effects, *Mon. Weather Review,* 107, 227-237, 1979.

L'vovich, M. I. and S. P. Ovtchinnikov, *Physical-Geographical Atlas of the World* (in Russian), State Geodetic Commission, Moscow, 1963.

Maugh, T. H. II, Acid rain's effects on people assessed," *Science,* 226, 1408-1410, 1984.

Mitchell, J. F. B., The hydrological cycle as simulated by an atmospheric general circulation model, 429-446, in *Variations in the Global Water Budget,* edited by A. Street-Perrott and M. Beran, D. Reidel, Boston, 1983.

Myers, N., *Tthe Sinking Ark,* 307 pp. Pergamon Press, N.Y., 1979.

NRC U.S. National Committee for an International Geosphere-Biosphere Program, *Global Change in the Geosphere-Biosphere,* National Academy of Sciences, Washington, D.C. (in press), 1982.

Perrier, A., Land surface processes: vegetation, in *Land Surface Processes in Atmospheric General Circulation Models,* edited by P. S. Eagleson, Cambridge University Press, 1982.

Phillips, J., in *Fire and Ecosystems,* T. T. Kozlowski and E. E. Ahlgren, editors, Academic Press, N.Y., 435-481, 1974.

Sagan, C., O. B. Toon, and J. B. Pollack, "Anthropogenic albedo changes and the Earth's climate, *Science,* 206, 1363-1368, 1979.

Salati, E. and P. Vose, The Amazon basin: a system in equilibrium, *Science,* 225(4658), 129-138, 1984.

Schutz, C. and W. Gates, Global climatic data for surface, 800 mb., 400mb, Reps. R-915-ARPA, R-915/1-ARPA, R-1029-ARPA, Rand Corp., Santa Monica, 1971.

Shukla, J., Effect of Arabian Sea surface temperature anomaly on Indian summer monsoon: a numerical experiment with the GFDL Model," *J. Atmos. Sci.*, 32, 503-511, 1975.

Shukla, J. and Y. Mintz, Influence of Land-Surface evapotranspiration on the Earth's climate," *Science,* 215, 1498-1501, 1982.

Shuttleworth, W. J., Evaporation models in the global water budget, 147-171, in *Variations in the Global Water Budget,* edited by A. Street-Perrott and M. Beran, D. Reidel, Boston, 1983.

Thornthwaite, C. W. and J. R. Mather, The water balance, *Publications in Climatology,* 8(1), Laboratory of Climatology, Centerton, N. J., 86 pp, 1955.

Walsh, J. E., Snow cover and atmospheric variability, *American Scientist,* 72, 50-57, 1984.

Webb, T., J. Kutzbach, H. Wright, E. Cushing, A. Street-Perrott and W. Ruddiman, unpublished data from the Cooperative Holocene Mapping Project (COHMAP).

Webster, P. J., Seasonality in the local and remote atmospheric response to sea surface temperature, *J. Atmos. Sci.*, 39, 41-52, 1982.

World Meteorological Organization, Report of the meeting of experts on the design of a pilot atmospheric-hydrological experiment for the WCRP, WMO-WCP Publication No. 76, Geneva, 1984.

PETER S. EAGLESON

Massachusetts Institute of Technology, Cambridge, MA 02139

Peter S. Eagleson was born in Philadelphia, Pennsylvania, where he had his primary education. Following graduation from Cheltenham High School in Elkins Park, PA, he entered Lehigh University in Bethlehem, PA, where he obtained B.S. and M.S. degrees in Civil Engineering. Peter Eagleson entered MIT in 1952, where he joined the Department of Civil Engineering. In 1956 he received the Sc.D. degree and since 1965 has served as Professor in that department. From 1970 to 1975, Peter Eagleson was Head of the Department of Civil Engineering at M.I.T.

Dr. Eagleson's major scientific and professional interest is in theoretical hydrology. He is the author of a fundamental text in *Dynamic Hydrology,* and the editor of a monograph on *Land Surface Processes in Atmospheric General Circulation Models.* His honors and awards include the Desmond Fitzgerald Medal of the B.S.C.E. (1959), the Research Prize of A.S.C.E. (1963), the Clemens Herschel Prize of the B.S.C.E (1965), the John R. Freeman Memorial Lectureship of the B.S.C.E. Section /A.S.C.E. (1977), and the Robert Horton Award of the A.G.U. (1979).

Peter Eagleson has served on a number of national and international scientific committees, including the Review Panel on State of the Art of Remote Sensing in Earth Sciences (National Academy of Sciences), the NASA Science Working Group for Land-Related Global Habitability, the Water Science and Technology Board of the National Research Council. He chaired the Working Group of Land Surface Processes in the Global Atmospheric Research Program of the World Meterological Organization. Dr. Eagleson is an active member of several professional societies that include the American Society of Civil Engineers, Boston Society of Civil Engineers, International Association for Hydraulic Research and the American Geophysical Union. He became a Fellow of the AGU in 1973, president of the Hydrology Section (1980-1984), and is currently President of the American Geophysical Union.

In 1982, Peter Eagleson was elected to the National Academy of Engineering.

FOREWORD

In the foreword to the Fourth Kisiel Memorial Lecture we defined hydrology as the science that studies the interaction on planet Earth between hydrosphere on the one hand, and lithosphere, biosphere and atmosphere on the other. The purpose of the study is to understand nature, discovering its laws. Since water is all-pervasive on this planet — even in the driest desert there is water in vapor form in the soil pore space — we may ask whether the study of water vapor in arid soils is a chapter in hydrology. The answer does not seem to be obvious since pedology could claim it to be within the purview of its concern, which indicates that Science is a continuum that we have partitioned mostly for our convenience creating some overlaps in the process.

Emerging from this area of possible contention between two earth sciences, we may ask the further question whether the study of water as a fluid with specific physical characteristics would get us any closer to an understanding of hydrologic phenomena. The answer is affirmative only for some hydrologic phenomena, such as flow in porous media.

These two examples should suffice to suggest that different hydrologic laws may be discovered at different scales: scales of length and scales of time. Combining length and time in an appropriate fashion we obtain velocities. Indeed, the movement of water in nature (what we call "the hydrological cycle") occurs at different velocities in different media: considerably slower below land surface than on it. The practical import of these different scales of velocities is our inability, so far, to produce a wholly satisfactory mathematical representation of regional water resources consisting of surface streams and groundwater aquifers for the purpose of deriving development policies and operating rules that would meet user demands at an acceptable cost.

Hence, problems of scale in hydrology are of considerable importance, theoretical as well as practical. The Fifth Kisiel Memorial Lecture surveys the scene and presents to us a state-of-the-art summary of these important issues. What is more important, perhaps, is that it brings us one step closer to bridging the scale gaps *en route* to a more comprehensive representation of hydrologic phenomena.

Nathan Buras

SCALE PROBLEMS IN HYDROLOGY

JAMES C.I. DOOGE

1. ORDERS OF MAGNITUDE

1.1 Choice of topic

The fact that the four previous Kisiel Memorial Lecturers all mentioned the dual role of Hydrology as a scientific discipline and as a basis for informed decision-making on important practical problems reflects the importance of this question for the Hydrologist. I was particularly struck by the analogy used by Mike Fiering who said (Fiering 1984: 3):

> As a scientist,
> I would rather be right;
> but as an engineer concerned with water,
> I would rather be president.

Reading these words I was reminded of the words of the English writer and politician Christopher Hollis in one of his early books (Hollis 1927: 110):

> "Clay, as every American school-boy is taught, would have rather been right than have been President. When the hard experience of life showed him that it was quite impossible to be either, it would perhaps be unreasonable to expect him to have displayed equal scruples about a mere Secretaryship of State."

The hard experience of hydrological life may well persuade the hydrologist that it is impossible either to develop a true science of hydrology or to provide a reliable basis for decision making in water resources planning. I hope, however, that no hydrologist ever has to compromise his principles in order to obtain the rank of Secretary of a State or its equivalent. Chester Kisiel strove mightily to harmonize these two objectives and in doing so gave guidelines in intellectual integrity to his colleagues who had known him as well as to a younger generation of hydrologists who know only his writings.

There does not appear to be any fundamental theorem or set of theorems in hydrology that can serve as a master key to unlock the secrets that govern the occurrence and movement of water on this planet. Given the state of scientific hydrology, it is unlikely that even the undergraduate students of today will survive until a time when such theorem or theorems are available or when a consistent and scientifically based corpus of knowledge capable of predicting all types of hydrological phenomena is available. Since we are unlikely to discover universal hydrologic laws, alternative strategies must be adopted in order to secure a scientific foundation for decision making in regard to the problems of the management of water resources which are of such vital economic and social importance to mankind.

My general topic today is the discussion of the research strategy based on the hypothesis that the best route to reliable and useful hydrologic knowledge may be through a recognition that different laws may be discovered at different scales and the appropriate set of laws then chosen in the light of the scale and type of problem to be solved. In some cases it may be possible to establish links between the formulation of hydrologic relationships at two adjacent scales and thus increase the logical strength of both the theoretical constructs and at the same time obtain more reliable parameter estimates as a basis for decision making. In other cases, the link between the mathematical formulation and the parameter values at adjacent levels of analysis may be so complex that no relationship between them can be established for many scientific generations to come.

Such a topic commends itself to me for two reasons. Firstly, I believe that the discussion of this topic is timely and necessary if advances are to be made, either in hydrological theory or hydrological practice. Secondly, and appropriately on the occasion of this Memorial Lecture because it is typical of the topics which were the subject of informal discussions among those mathematically-aware hydrologists who, a couple of decades ago, transformed themselves from a small and invisible college of correspondents into the I.A.H.S. Committee on Mathematical Models in Hydrology and ultimately into the International Commission on Water Resource Systems. Among that group and active in those informal discussions was Chester Kisiel. In such discussions and in his writings, Chester showed a keen awareness of the need to synthesize the deterministic and stochastic approaches and of the importance of the question of scale in this connection. I would like to think therefore that the topic is one in which he would have taken a lively interest and that the emerging debate on the subject is one to which he would have contributed if he were still with us today.

Recalling Chester's zeal in directing colleagues and students to original sources and in encouraging them to further reading on aspects of the topic that they found of particular interest, I have included in this written version of the Memorial Lecture a number of references that I found interesting and useful in the preparation of the Lecture. Needless to say, I would greatly welcome any reaction to the ideas put forward.

1.2 Subhuman and superhuman scales

The problem of scale does not arise only in hydrology. It poses difficulties for almost every science and is a key factor in our perception of the Universe of which we are a part. In this connection I would refer you to the film (Eames, 1977), and the book (Morrison and Morrison, 1982) which share the common title: "Powers of Ten." These both present "the relative size of things in the universe and the effect of adding another zero" (Morrison and Morrison 1982: 19). The film and book contain visual images ranging from a scale of 10^{25} meters showing galaxies as tiny clusters in empty space by steps corresponding to the conventional order of magnitude of ten to a scale of 10^{-15} meters at which we have an abstract visualization of the proton and its quarks. These pictures vary enormously as we move either outwards or inwards in terms of scale.

A remove of three orders of magnitude above or below the one meter scale of the human body brings us either to the typical height of the planetary boundary layer which controls our immediate environment or to the typical scale of 1 mm which controls our individual organic life. A scaling of six orders of magnitude brings us either to the regional scale of ten degrees of latitude or to the electron microscope scale of the cell nucleus. Nine orders of magnitude away from the 1 meter base brings us either to the scale of the orbit of the single moon of our planet or in the other direction to the scale of the DNA molecule, the pattern of which has only recently been revealed to the curiosity of science. If we move to twelve orders of magnitude, then the outer scale is that of the orbit of Jupiter the first of the outer planets of the Solar System and the inner scale is that of the atomic nucleus. At fifteen orders of magnitude from the scale of one meter the picture is either that of a carpet of stars of which the sun is one or the pattern of the interior of a proton.

The pictures that we traverse as we move from the largest of the superhuman scales to the smallest of the subhuman scales or vice versa show an interesting alternation between images that reveal a distinct structure and those that appear as aggregates of a large number of random elements. The structures that appear at various scales both superhuman and subhuman show a marked similarity to the constructs of our human mathematics. Among the superhuman scales we can recognize at a scale of 10^{21} meters the spiral structure of the Milky Way, the galaxy of which the solar system is a small part; at this scale of 10^{21} meters our region of the Milky Way appears as a carpet of stars of varying brightness. At a scale of 10^{23} meters the Milky Way appears not as a spiral but as one of a number of bright spots in empty space. A scale of 10^{13} meters encompasses the orbits of all the planets of the solar system except Pluto with their elliptical almost co-planar orbits together with most of the orbit of the outermost planet Pluto which is exceptional for being smaller in size and not co-planar.

At the scales between 10^{-7} and 10^{-9} meters we can see the various structures of the vital DNA molecule including the famous double helix configuration. At lower scales there is no definite pattern among the electron clouds until at scales of 10^{-13} and 10^{-14} the picture shows the regular configuration of the atomic nucleus in the appropriate geometric form. At smaller scales again we are in the realm of quarks and other elementary particles whose regularities are still mysterious to us.

The range of orders of magnitude of significance in hydrology is substantially less than the wide range of 40 orders of magnitude discussed above. Nevertheless some guidance towards tackling the scale problem in hydrology can be gained from a consideration of how the problem of divergent scales is handled in the various sciences. As one of the geo-sciences, hydrology must look to the history and present status of physics for inspiration in regard to those aspects of its subject matter that depend directly on the transport and transformation of energy. At the same time, hydrology can with benefit look to the biological sciences particularly in their systems formulation for help in the understanding of the results of the adaptive processes of geomorphology.

If we study the physical sciences in their historical context, we will probably gain more useful knowledge than if we look only at the end result encapsulated in a textbook on modern physics. From the history of the developments that led to the Newtonian synthesis, we can appreciate that the imaginative thinking of Copernicus in the first half of the 16th century, the superb observational achievements of Tycho Brahe in the second half of the 16th century and the long and detailed analyses of Kepler in the first half of the 17th century were all preludes to Newton's formulation of the theory of Universal Gravitation in the second half of the 17th century. Though modern physics tells us that relativity requires the basic equations from which planetary orbits are derived to be non-linear, Newton was able to produce a linear approximation of high accuracy. The physical hypothesis, the mathematical technique, and the observational precision necessary to improve on the linear solution were not available for more than a hundred years after Newton. The particular features of the solar system which made it susceptible to linear analysis of a deterministic form have been well discussed by Weinberg (1975).

The determinism inherent in the physical sciences of the seventeenth century was challenged by the statistical approach of Maxwell and Boltzman in the 19th century (Brush 1983). A study of the history of the development of physics will save us both from the opinion that the rejection of determinism is a confession of ignorance and an abandonment of logical thinking and also save us from the exaggeration that there is nothing left of use in the old approach. The former error has been well described by Eddington (1935: 73) when he said:

> "The rejection of determinism is in no sense an abdication of scientific method. It is rather the fruition of a scientific method which has grown up under the shelter of old causal method and has now been found to have a wider range. It has greatly increased the precision of the mathematical theory of observed phenomena."

The latter exaggeration has been pointed out by Brush (1983: 95) when he says:

> "The new physics was most definitely a legitimate offspring of the old, though adolescence brought defiant claims of independence and denials that anything useful could be inherited from a supposedly stiff and stuffy parent."

Chester Kisiel was a person who reflected the open attitudes expressed in these two quotations and never indulged in sterile arguments about the relative merits or truth of parametric and stochastic hydrology.

Since most universities (unlike the University of Arizona) do not offer an undergraduate program in hydrology, most hydrologists come to the subject with inbuilt prejudices in regard to problem analysis derived from the nature of their undergraduate discipline or early post-graduate work. Some of them maintain these prejudices throughout their hydrologic careers. Hydrology has suffered much from the presumption that deterministic and stochastic methods are in some sense competitors for the same prize or that linear and non-linear approaches must always be antagonistic rivals.

1.3 Range of scales in hydrology

While we need not be concerned in hydrology with the complete range of scales from the quark to the galaxy discussed in the last section, nevertheless a considerable part of this range of scale variation is of significance in the search for hydrologic laws. The usual practice in scientific analysis is to classify phenomena in three groups and to label them as macro-scale, meso-scale and micro-scale. Such a tripartite division is insufficient for a thorough discussion of the various levels of analysis required in hydrology. Accordingly, each of these three divisions has been further subdivided into three and the nine categories and the approximate length scales involved are indicated in Table A.

The largest of the three macro-scales is the planetary scale which is taken as 10^7 meters which was the basis of the metric system when established in 1793, being the current estimate of the distance from the pole to the equator, measured on a meridian passing through Paris. This represents approximately the scale at which it is appropriate to discuss the evolution of the hydrosphere of our planet and climatic changes on a global scale. The central macro-scale or continental scale is taken as 10^6 meters and corresponds to a few mesh lengths of a general circulation model of the earth's atmosphere as used for climate studies. The smallest of the three macro-scales is that of 10^5 meters representing the largest scale at which detailed measurements are currently being planned in the first of the series of hydrologic atmospheric pilot experiments (HAPEX) which is now in operation (WCRP 1985). It is the scale of a large catchment.

The variation in the meso-scale is from the higher meso-scale corresponding to a catchment area of approximately 100 km^2 through a subcatchment on one order of magnitude lower to a catchment module with a somewhat higher degree of homogeneity characteristic of a one hectare field or similar module in a subcatchment. These three meso-scales represent typical scales at which hydrologic data is available in relative abundance.

When we come to the micro-scales, at which hydrology is linked through hydraulics and fluid mechanics to the physical sciences, we have a greater spacing between the three levels that have to be considered. Table A suggests that these can be taken for discussion purposes as some representative elementary volume as used in hydrogeology and the study of the flow through porous media generally. This might be taken as the intermediate value 10^{-2} meters corresponding to typical soil but would be smaller for clays and larger for karsts. The scale below which the continuum hypothesis, which is the basis of ordinary fluid mechanics, is no longer appropriate is often taken as about 10^{-5} m. The lowest of the micro-scales is that corresponding to a cluster of water molecules and is taken for the present purpose as 10^{-8} m, i.e. a hundredth of a micron.

TABLE A

SPATIAL SCALES IN HYDROLOGY

CLASS	SYSTEM	Typical Length (Meters)
MACRO	PLANETARY CONTINENTAL LARGE CATCHMENT	10^7 10^6 10^5
MESO	SMALL CATCHMENT SUB-CATCHMENT MODULE	10^4 10^3 10^2
MICRO	REPRESENTATIVE ELEMENTARY VOLUME CONTINUUM POINT MOLECULAR CLUSTER	10^{-2} 10^{-5} 10^{-8}

It is obvious that the classification suggested in Table A as the basis for discussion is a somewhat arbitrary discretization of a continuous range of scales covering phenomena of widely different types. It is felt, however, that such a nine-fold categorization of scales represents a useful basis for the discussion of scale problems in hydrology. The ultimate aim of scientific hydrology would be to discover universal laws from which deductions could be made at any of these scales. It is clear that such an accomplishment is still beyond our grasp. What is intended in the present survey is a discussion of the relationships between these nine scales as a contribution to the discussion of the present position and of the prospects of advancement.

In the description of spatial scale, both in this section and the previous one, the various scales have been listed in descending order from the largest of the superhuman scales to the smallest of the subhuman scales and from the planetary to the molecular. In discussing the types of analysis appropriate at each level and the linkages between the levels, it is more convenient and logical to discuss the various scales in ascending order from the smallest to the largest. Accordingly, the remaining section of this Lecture will deal with the nine spatial scales listed on Table A, starting from the micro-scale of a cluster of water molecules at 10^{-8} meter and working up to the macro-scale of the entire planet at 10^{7} meters. When this is done, significant differences in approach become apparent over this range of fifteen orders of magnitude. In a broad sense it can be said that at the micro-scale we are concerned with the application to hydrologic processes of the standard methods used in the physical sciences; at the meso-scale we are concerned with the dynamic behavior of systems which are intermediate in size and nature between simple mechanisms and large aggregates of random elements; at the macro-scale we become more concerned with the conditions for the equilibrium of complex systems involving a number of feedbacks and interactions.

While the above discussion is structured on the basis of the range of spatial scales, the question of the range of time scales involved is also highly relevant. Schumm and Lichtey (1965) have already proposed a tripartite set of scales of relevance in the study of geomorphic processes. They describe cyclic time as the length of time (millions of years) required for an erosion cycle to evolve through dynamic equilibrium between geology, initial relief and climate; graded time as the shorter span (hundreds of years) during which a river could develop a graded profile representing a steady-state equilibrium between climate, vegetation, relief and hydrology; and steady time as a period (perhaps a month or less) in which there is steady state equilibrium and a fixed relationship between the discharge of water and of sediment and the channel characteristics at a given section or in a given reach of channel.

For our purposes we will again use a nine-fold classification descending from a scale appropriate to the evolution of the solar system and of the earth's first primitive atmosphere to a scale appropriate to the behavior of water molecules. The result of attempting such a classification is shown on Table B together with the approximate order of magnitude involved. As in the case of space scales the main concentration in hydrology is on phenomena at the meso-scale where the concern is with variations from year to year, or seasonal variation from month to month, or with more rapid variations from day to day. These are natural time scales corresponding to the three body system of sun, earth and moon and were obvious as natural time scales to our remote ancestors.

Above these meso-scales of immediate interest are such periodic phenomena as the sun-spot cycle of eleven years, the

Milankovitch cycle of 100,000 years due to slight variations in the eccentricity of earth's orbit about the sun, and the ultimate cosmological time scale of stellar evolution which is the order of magnitude of a thousand million years.

In the physical study of hydrological processes we may have to deal with time scales shorter than the lowest meso-scale of 1 day. In the case of the results of experiments on small plots we may at times measure changes in runoff in seconds which is almost five orders of magnitude smaller than days. The time scale in continuum mechanics corresponding to the space scale of 10^{-5} m may be taken arbitrarily as 10^{-6} seconds. Recent molecular models capable of explaining the anomalous properties of water indicate a time scale of 10^{-13} seconds for the continual formation and breakdown of clusters of water molecules (Dooge 1983). It is interesting to note that some of these anomalous properties determined at the molecular scale are significant at the planetary scale because they enhance the performance of water as a climate modifier.

TABLE B

TIME SCALES IN HYDROLOGY

CLASS	TYPE	ORDER OF MAGNITUDE
MACRO	STELLAR EVOLUTION ORBITAL ECCENTRICITY SUN SPOTS	10^9 YEARS 10^5 YEARS 10 YEARS
MESO	EARTH ORBIT MOON ORBIT EARTH ROTATION	1 YEAR 1 MONTH 1 DAY
MICRO	EXPERIMENTAL PLOT CONTINUUM POINT WATER CLUSTER	1 SECOND 10^{-6} SECOND 10^{-13} SECOND

1.4 Role of dimensional analysis

The notable successes of dimensional analysis in fluid mechanics and in hydraulics might lead one to believe that the techniques necessary to span the range of variation in scale in hydrology could be founded on similar considerations. It might appear that all that was required was to find a hydrologic analogue to the Reynolds number which enables a chaotic mass of data on the flow of various fluids through pipes of various size and roughness to be reduced to a single curve or an analogue to the Froude number which acts so efficiently as the basis for modelling in the study of hydraulic structures. There are a number of indications that the difficulties in the case of hydrology are more fundamental than a mere ignorance of the basic law from which such a dimensionless number could be derived by inspectional analysis, in the same way as the Reynolds number or the Froude number can be derived from Newton's second law.

In the type of phenomena dealt with in hydrology, the nature of dimension may itself be different from the concept of integral dimension appropriate in fluid mechanics and hydraulics (Mandelbrot 1977, 1983). It is interesting that the key question relating the idea of fractional dimension to geophysical data ("What is the length of a coastline?") was posed by L. F. Richardson who worked on both macro-scale problems and micro-scale problems. In 1922 he wrote the classical work on the macro-scale problem of the numerical analysis of the general circulation of the atmosphere and in 1926 published an important paper on the application of self-similarity to the micro-scale problem of turbulence. This matter of fractal dimension is of prime importance in hydrology because the same difficulty arises in regard to the definition of such key parameters as the drainage density which involves not only accurate determination of the length of the streams of the drainage network but also a clear definition of the catchment area. Mandelbrot (1977, 1983) has thoroughly and elegantly discussed the nature of this problem at a wide variety of scales.

Another difficulty in seeking a single key to the unlocking of hydrological similitude is the fact that the similarity of behavior at the meso-scale in space and time may depend on evolutionary processes at macro-scale in both space and time. In this connection, the experience of the use of similitude in biology has some lessons for the hydrologist. It has long been recognized that the rate of metabolism among animals could not be proportional to their volume. The reasonable hypothesis was then put forward that the rate of metabolism would be proportional to oxygen uptake and hence proportional to a surface area rather than a volume. The resulting variation of the rate of metabolism with the two-thirds power of the weight was empirically confirmed for variations in weight among members of a single species but when mammals and birds of different species were compared there was an almost exact relationship between the rate of metabolism and the three-fourths power of the weight (Kleiber 1932) as shown on figure 1 (Benedict 1938).

The empirical result known as Kleiber's Law can be explained by the hypothesis that on the time scale of evolution the proportions of the animal body are determined by the requirement that the legs of the animal should not be subject to buckling during running either in pursuit of prey or in fleeing from a predator. The assumption of elastic similarity (D'Arcy Thompson 1917: 988, McMahon 1984: 258–265) based on Euler's Law for the buckling of columns and of a constant strength of bone and a constant ratio of muscle cross-section to bone cross-section lead to the prediction of Kleiber's Law (McMahon 1984: 278–280). It is interesting to note McMahon's conclusion after applying three candidate similitude laws to a number of biological phenomena when he says (McMahon 1984: 293):

> "Perhaps the most realistic judgment would be that no single principle unifies all observations on animal scaling: . . . The arguments of this chapter conclude with the contention that elastic similarity is not a perfect theory of animal scaling, only a useful one."

It is noteworthy that Kleiber's Law, which has good empirical support (Kleiber 1932, Benedict 1938), gives a scale law close to but clearly different from that based on geometrical similitude.

It may well be that in the case of flood hydrology and similar problems the relationship between flood run-off and catchment area and catchment slope may be different but not very different from the values indicated by the direct application of the Froude criteria as used in designing scale models in hydraulics (Dooge 1986a). The situation in hydrology is worse in that we do not even have a useful theory of hydrologic similitude that can rival Kleiber's Law which governs the size variation from the mouse to the elephant. The derivation of such new laws of similitude would be extremely difficult because the quality of the data and the links between the variables make regression analysis a very blunt instrument in the absence of a rational hypothesis that can be systematically tested.

2. MICRO-SCALE MODELS IN HYDROLOGY

2.1 Water as an isotropic continuum

Hydrology is concerned both with the occurrence and the movement of water. Most of the properties of water are anomalous in one way or another compared with the corresponding chemical compounds involving the neighbors of oxygen in the periodic table. If we compare water, which is an oxygen anhydride, with the other group VI anhydrides we would expect water to occur in liquid form in a range from about −65°C to about −85°C instead of from 0°C to 100°C. This and other anomalous properties of water are explained by the physical chemist on the basis of the polarity of the water molecule, i.e. the fact that the electrical charges of the component atoms do not share a common center of action and hence produce a dipole moment even though they are in electrical balance (Speakman 1966, Dooge 1983). Neutron diffraction studies show that the oxygen atoms in water are linked through a shared hydrogen atom thus giving rise to hydrogen bonding which is supplementary to the ordinary covalent bonding and about one tenth as strong. This extra bonding enables ice to maintain its solid state until the temperature rises above zero on the Celsius scale compared with the expected value of about −85°C and for liquid water to remain in that condition until it reaches 100°C compared with the expected value of −65°C.

When we turn to consideration of the movement of liquid water rather than its occurrence we abandon completely the assumption that water is highly non-isotropic, which was the basis of the explanation for the occurrence of water in liquid form at the temperatures which we encounter over most of our planet. Not only is the molecular structure of water completely ignored and replaced by the continuum hypothesis, but this continuum is assumed to be isotropic in flat contradiction to the basic assumption at the molecular scale. It would appear that we are prepared to assume that water, which appears to consist of highly non-isotropic molecules forming clusters at a scale of 10^{-8} m will, under all conditions of flow, display completely isotropic behavior on the scale of 10^{-5} m by representing the aggregation of these molecular clusters into a volume of this scale which is represented by a point in continuum mechanics. While the direct linking of the parameters of these two scales is possible for the case of gases where the individual molecules show no structure and in the case of solids where the individual molecules are strictly structured, such a connection cannot be made in the case of a liquid as complex as water where the molecules are loosely structured.

What then is the justification of the Navier-Stokes equation which is the basic equation of fluid motion derived by the methods of continuum mechanics? It rests largely on the fact that the assumption of water as an isotropic continuum simplifies the general problem to the extent that the movement of water at a continuum point can be represented in terms of two material parameters only and in terms of a single parameter if compressibility is neglected

(Dooge 1983). The degree of simplification achieved by the single assumption of isotropy is remarkable.

Even when we assume that the local shear tensor is a linear function of the local deformation tensor (i.e. assuming that liquid water is a Newtonian fluid), the resulting relationship:

$$\tau_{ij} = C_{ijrs}(\rho,\theta).D_{rs} \tag{1}$$

is a complex one. In this equation both the local shear τ_{ij} and the local deformation D_{rs} are second order tensors and hence the coefficient linking these two tensors C_{ijrs} (which is a function only of the thermodynamic variables) must be a fourth-order tensor. In general a fourth-order tensor contains 81 elements but since the two second-order tensors in equation 1 are symmetric, the linking fourth-order tensor C_{ijrs} will also be symmetric and hence the number of independent elements required to describe the behavior of the fluid will be reduced from 81 to 36. The formidable problem presented by such a complexity can be reduced to manageable proportions by making the single assumption that the fluid under discussion is isotropic. When this is done we can replace equation (1) for a general Newtonian fluid by:

$$\tau_{ij} = \lambda D_{rr}.\partial_{ij} + 2\mu D_{ij} \tag{2}$$

which describes the behavior of an isotropic Newtonian fluid. In equation (2) the relationship between the shear and the local deformation can be explained in terms of the bulk viscosity λ and the dynamic viscosity μ. If compressibility is neglected then the divergence D_{rr} is zero and the dynamic viscosity μ is the only material parameter in the constitutive equation.

If the direct stress p and the shear stress τ as defined by equation (2) are substituted in the equation for the conservation of linear momentum we obtain the Navier-Stokes equations (written in tensor notation with the summation convention of repeated indices) as:

$$\rho\left(\frac{\partial u_i}{\partial t} + \frac{u_j \partial u_i}{\partial x_j}\right) = F_i - \frac{\partial p}{\partial x_i} + \mu \frac{\partial^2 u_i}{\partial x_j \partial x_j} \tag{3}$$

where F_i is the i-component of the body force. The apparent contradiction involved in these two approaches should not deter us from using either the non-isotropic model of physical chemistry or the isotropic model of continuum mechanics when dealing with problems appropriate to it. Rather should we accept the lesson that problems of differing scale can provide accurate predictions and promote understanding of problems appropriate to each of the two scales through the use of apparently conflicting approaches.

What value to us is the Navier-Stokes equation reached through a string of assumptions, which has been shortened and summarized above, in the solution of real problems of the movement of water? The solution of the Navier-Stokes equation, even for simple boundary conditions, is difficult because the second term on the left hand side of equation (3) involves a non-linearity. In order to apply the Navier-Stokes equations to hydrologic processes it is necessary to be able to use it in the analysis of flow through porous media both saturated and unsaturated and of free surface flow both laminar and turbulent. In neither of these two types of problem is it possible to use the Navier-Stokes equation in the basic form of equation (3) above. In order to study flow through porous media it is necessary to derive the Darcy equation which is applicable to a higher scale than that of the continuum point at which the Navier-Stokes equations are applicable and to establish parameter values at this higher scale appropriate to a representative elementary volume of the porous medium. In the case of flow in open channels, it is necessary to establish a new formulation at a higher scale than the continuum point because the flow conditions become impossible to analyze at the continuum point scale after the onset of turbulence. In both cases simplifications are introduced in practice in order to understand the phenomena involved and to enable useful solutions to be found to real problems.

2.2 Creeping flow in porous media

In analyzing flow through porous media in hydrology, Darcy's equation relating an average face velocity to the potential gradient (Darcy 1856: 305–310, Hubbert 1940: 785–803, Scheidegger 1960: 68–90) is much more useful than the Navier-Stokes equation applicable to a continuum point. Darcy's Law only applies to creeping flow, i.e. to slow flows for which the Reynolds number is less than unity. At higher Reynolds number we have a succession of types of flow including steady non-linear flow with an inertial core, unsteady laminar flow and finally a highly unsteady chaotic flow resembling turbulence (Dybbs and Edwards 1984). For such creeping flows the Navier-Stokes equation for a continuum point can be simplified to:

$$\mu \frac{\partial^2 u_i}{\partial x_j \partial x_j} = F_i - \frac{\partial p}{\partial x_i} \tag{4}$$

but even in this simplified form can only be solved for simple boundary conditions.

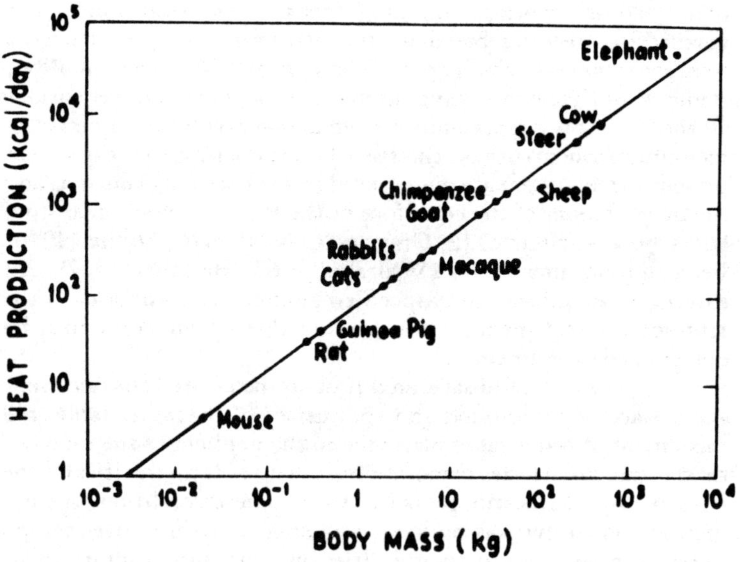

Figure 1. Scale effect in animals.

In order to derive the Darcy equation it is necessary to integrate equation (4) not only over the cross-section of each individual pore but to integrate a second time the mean velocity in each individual pore across the representative area which is appropriate to the scale at which the analysis is being made. If this can be accomplished then the average face of velocity which results from this double integration is given by:

$$\bar{u} = \frac{k_i}{\mu} \left(F_i - \frac{\partial p}{\partial x_i} \right) \tag{5}$$

where k_i is the intrinsic permeability of the solid matrix in the i-direction.

Three types of approach have been used in attempts to link the intrinsic permeability with the matrix properties of the porous medium; (a) modelling the porous medium by geometry simple enough to allow for the direct solution of the Navier-Stokes equations, (b) statistical models involving various assumptions in regard to the ensemble, the dynamics at the micro-scale and the statistical distribution, (c) the averaging at the Darcy scale of a representative elementary volume of the governing differential equations of the macro-scale by a logically consistent procedure. The class of models based on simplified geometry were reviewed by

Scheidegger (1960: 112–138) and by Dullien (1979). Scheidegger grouped such geometrical models into capillaric models, hydraulic radius theories, and drag theories. In contrast to these geometrical theories based on the deterministic approach are a number of statistical theories (Scheidegger 1965, Aranon 1966, Beran 1968). The averaging approach seeks to develop equations at the scale of a representative elementary volume in terms of measurable quantities on the basis of considering that the phases present in the system are represented by over-lapping continua and the development of the equations at the higher scales is a far from trivial process (Bear 1972, Gray and O'Neill 1976, Dullien 1979, Bear and Bachmat 1984, Whitaker 1967, Bachmat 1972). All three approaches lead to unspecified parameters at the scale of the representative elementary volume and these parameters must be determined experimentally.

In the case of unsaturated flow in the aerated zone between the surface of the ground and the surface of the water table, the movement of water takes place due to the gradient of the total potential which includes osmotic pressure as well as gravity and the negative hydrodynamic pressure. Under these conditions the hydraulic conductivity is no longer constant but varies over several orders of magnitude as the soil dries out from saturation to an air dry condition. The parametrization of the capillary potential, the negative hydro-dynamic pressure and of the unsaturated hydraulic conductivity in terms of moisture content make the problem more complicated than that of saturated flow.

2.3 Turbulent flow in open channels

In principle the application of the Navier-Stokes equation to turbulent flow is quite impracticable even with the aid of the largest computer available or indeed imaginable. The first step in reducing this problem to a more manageable form is to average the Navier-Stokes equation as given by equation (3) above over a time period during which the average of the turbulent fluctuation from the mean local flow may be assumed to be zero. When this is done, the non-linear term on the left-hand side of equation (3) causes further difficulty since the average equation will be:

$$\rho\left(\frac{\partial u_i}{\partial t} + \frac{\bar{u}_j \partial \bar{u}_i}{\partial x_j} + \frac{\partial(u_i' u_j')}{\partial x_j} \right) = F_i - \frac{\partial p}{\partial x_i} - \mu \frac{\partial^2 \bar{u}_i}{\partial x_j \partial x_j} \tag{6}$$

where $\bar{u}_i$ and $\bar{u}_j$ are mean velocities and u_i' and u_j' are deviations from this mean velocity at the continuum point scale of the Navier-Stokes equation.

In practical cases of interest, the cross-product term on the left-hand side of equation (6) is known as a Reynolds stress and is at least an order of magnitude greater than the viscous stresses on the right-hand side of the equation due to the gradient in the mean velocity. These Reynolds "stresses" which arise from momentum interchange at the micro-scale must be linked to the mean variables flow in order to obtain a set of equations capable of yielding a solution. Such turbulence closure models may be of differing degrees of complexity including; (a) algebraic relationships such as the Prandtl mixing length (Prandtl 1925), (b) a single differential equation for some property of the turbulence such as the transport of turbulent kinetic energy (Prandtl 1945) or (c) a set of differential equations involving a number of variables, e.g. the kinetic energy of turbulence and a characteristic energy frequency (Kolmogorov 1942). Since the advent of computers a number of proposals have been made involving considerably more complex turbulence models (Launder and Spalding 1982, Rodi 1980).

The Reynolds equations given by equation (6) above are still three-dimensional in form. A second step towards finding a manageable solution that will be useful in practice is to reduce them to one-dimensional form. This can be done by averaging over the cross-section but this will involve the appearance of dispersion terms due to the non-uniformity of the flow in a vertical direction. Instead of operating from conditions at a point it is possible to formulate equations required in hydrologic analysis directly for either a representative elementary volume or for a complete cross-section.

In the case of open channel flow, the one-dimensional equation was derived by Saint-Venant (1871) by considering a slice of water of infinitesimal thickness under acceleration due to gravity and deceleration due to boundary shear. Saint-Venant (1871) suggested the equation:

$$\frac{\partial \bar{u}}{\partial t} + \bar{u}\frac{\partial \bar{u}}{\partial x} + \frac{\tau_0}{\rho R} + g\frac{\partial z}{\partial x} = 0 \qquad (7)$$

where τ_0 is the average boundary shear, R is the hydraulic radius and z is the elevation of the water surface. As in the case of the Reynolds equation, a further closure equation is required in order to link the average boundary shear to the mean variables of flow. The Manning formula represents such an empirical closure equation based on an analysis of the measurements of open channel flow by Bazin.

In the case of flow in open channels, a good deal of insight into the movement of flood waves in rivers can be obtained through the study of the solution of linearized Saint-Venant equa-

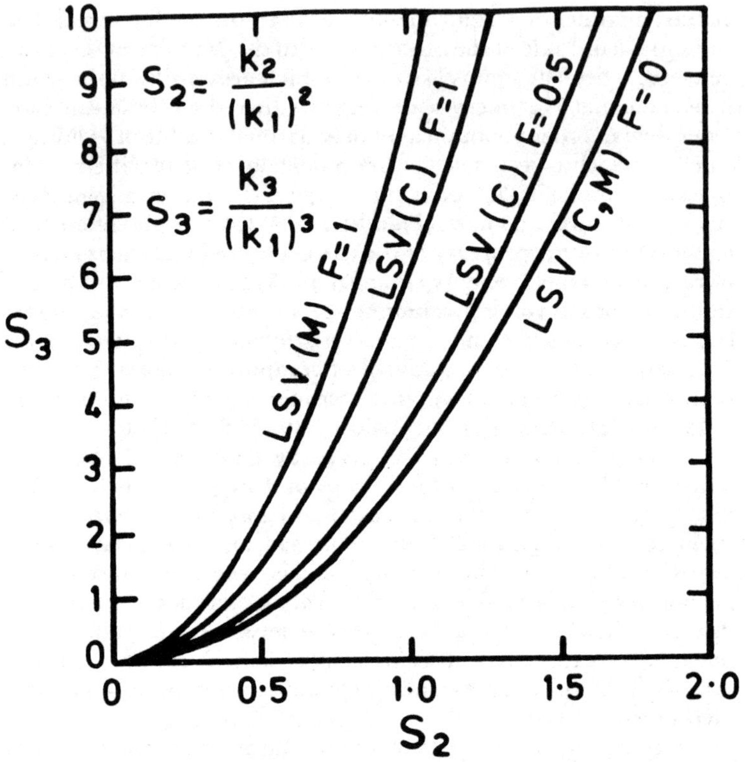

Figure 2. Shape factors based on lag.

tion (Dooge and Harley, 1967, Dooge 1973, Dooge et al. 1986). Thus it can be shown that the second central moment of the linear channel response is independent of the reference discharge about which the linear perturbations are taken. Since the third and higher moments have a diminishing effect on the shape of the linear channel response, it can be conjectured that the main error of linearization will be in the first moment, i.e. the lag of the channel response.

The shape of the linear channel response, as in the case of any other shape, can be conveniently studied through the use of dimensionless moments (Nash 1959, Dooge 1973, Strupczewski and Kundzewicz 1980). Dimensionless moments based on the first moment have been used in hydrology for catchment response by Nash (1959) and for channel response by Dooge and Harley

(1967). Figure 2 shows such a plotting of the dimensionless third moment based on the lag against the dimensionless second moment based on the lag for both Chezy and Manning friction in a wide rectangular channel. It will be noted that the curves for Froude number $F = 0$ and a Froude number $F = 0.5$ plot very closely together thus indicating very little change in shape over this range.

If it is accepted that there will be a variation of lag due to non-linearity then the effect of this can be eliminated by using dimensionless moments based on the second moment and allowing the non-linear effect to be absorbed in the first moment (Strupczewski and Kundzewicz 1980, Dooge 1980). Figure 3 shows for the same case as figure 2 by plotting the dimensionless fourth moment based on the second moment and the dimensionless third moment based on the second moment. It is seen from figure 3 that there is far less variation over the whole range of the Froude number from $F = 0$ to $F = 1$ and that at the two ends of the range there is no difference between the shapes for the different friction formulae. This suggests that, once non-linearity has been allowed for in the lag of the channel response, there will be little variation in shape.

It is clear that the physical equations used in hydrology either for free-surface flow or for flow through porous media are not directly related to physics as understood by the physicist or the physical chemist. The equations of physical hydrology represent several levels of parametrization from a molecular scale to the scale of a continuum point and on to the scale of a representative elementary volume or a finite cross-section of a channel. This raises the question whether an attempt to parametrize to still higher scales would be to extend the cantilever of scientific reasoning further than would be appropriate to its strength at the level of validated theory.

3. MESO-SCALE MODELS IN HYDROLOGY

3.1 Simple solutions for homogeneous modules

If one wishes to construct a meso-scale model of catchment response, then it is desirable that the individual modules incorporated in the total catchment model should be as simple as possible not just for ease of operation but also in order to make the parameters determined during calibration at the meso-scale more reliable. In addition, any attempt to allow for spatial variation in parameters must be guided by some knowledge of the sensitivity of the resulting predictions to various forms of simplification of the physically based equations such as reduction to one-dimensional form or linearization or simplification of the equation itself. Com-

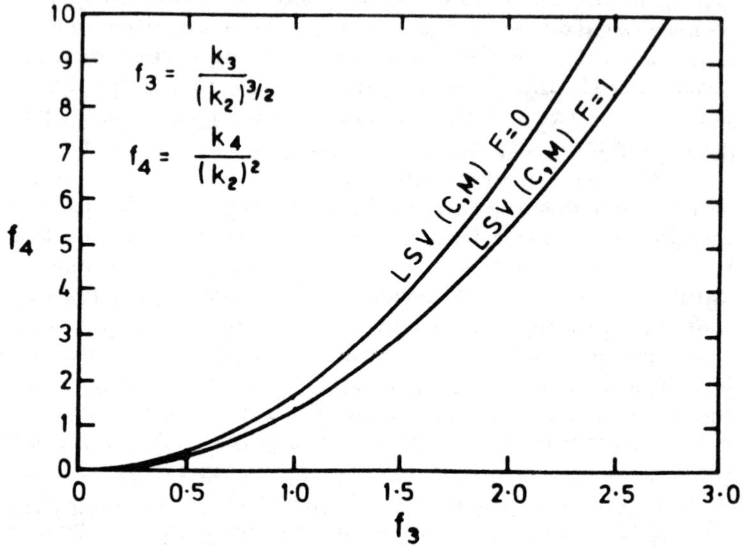

Figure 3. Shape factors based on variance.

parisons are given in the present section between one-dimensional and two-dimensional saturated steady ground water flow, between two solutions for ponded infiltration into the unsaturated zone based on widely different assumptions of soil properties, and between the linear channel responses obtained when different numbers of terms are retained in the linearized Saint-Venant equation.

It is customary when dealing with groundwater flow to use the Dupuit-Forschheimer assumptions which reduce the problem from two-dimensional flow to one-dimensional flow by neglecting vertical accelerations and assuming the flow to be essentially horizontal. It is well known that the profile of the groundwater surface obtained using such an assumption will be in error particularly in regions where the phreatic surface increases rapidly (Bear 1972). However, it was shown by Charnyi (1951) for the case of steady flow through an embankment that the actual steady discharge given by the Dupuit-Forschheimer solution was identical to the discharge given by the full two-dimensional solution. Numerov extended this proof to the case of seepage in a rectangular body of ground in the presence of vertical infiltration (Aravin and Numerov 1953, 1965: 88–90). Since we are concerned in hydrology with the groundwater outflow into a drainage trench or a field drain, the error in the profile does not concern us as long as the discharges are correct. Accordingly, the effect of the one-dimensional approximation on hydrological prediction is negligible for the cases cited.

A similar result can be obtained in regard to the method of linearization used in the case of unsteady groundwater flow. The one-dimensional equation for such flow is:

$$\frac{\partial}{\partial x}\left(K.h\frac{\partial h}{\partial x}\right) + r(x,t) = f\frac{\partial h}{\partial t} \tag{8}$$

where h is the height of the water table above the horizontal impermeable layer, f is the drainable pore space (assumed to be constant), r(x,t) is the rate of recharge at the surface of the water table, and K is the saturated hydraulic conductivity. This equation can either be linearized in terms of h which involves an approximation of the first term on the left-hand side of the equation or linearized in terms of h^2 in which case the term on the right-hand side of the equation is approximated. The profile obtained by these two methods of linearization is quite different. Thus the steady state solution for linearization in h is a parabola whereas the steady state solution for linearization in h^2 is an ellipse. However, if the assumption made in the linearization is also used in the computation of the flow into a drainage ditch or a tile drain, then exactly the same expression is obtained in each case. Thus for given values of the parameters, once again a difference in profile is not reflected in a difference in the outflow.

The equation for one-dimensional vertical unsteady flow in the unsaturated zone is given by:

$$\frac{\partial}{\partial z}\left[D(c)\frac{\partial c}{\partial z}\right] + \frac{\partial K(c)}{\partial z} = \frac{\partial c}{\partial t} \tag{9}$$

where c is the unsaturated moisture content, z is the elevation measured vertically upwards, K(c) is the unsaturated hydraulic conductivity, and D(c) is the unsaturated hydraulic diffusivity defined by:

$$D(c) = K\frac{\partial h}{\partial c} \tag{10}$$

where h is the pressure head which is negative in an unsaturated soil. The classical infiltration problem for which a solution to equation (9) is sought is that of ponded infiltration of water through the soil surface when the initial moisture constant is the same at all levels and the surface is suddenly saturated. If it is assumed that the wetting front following infiltration of water through the surface is completely sharp (i.e. no diffusion) then it can be shown that the variation with time of the infiltration at the surface which is of hydrological interest is given by:

$$f_e(t) = \left[\frac{(c_{sat} - c_0)K_{sat} \cdot S_0}{2t} \right]^{1/2} \tag{11}$$

where $f_e(t)$ is the initial high rate infiltration, c_{sat} is the saturated moisture content, c_0 is the initial moisture content, K_{sat} is the saturated hydraulic conductivity and S_0 is the initial soil moisture suction. It has been remarked by Philip (1954) that this approach due to Green and Ampt (1911) is equivalent to taking the relationship between the hydrologic diffusivity D and the moisture content c as a Dirac delta-function.

If the completely opposite assumption is made that the hydrologic diffusivity is constant for all moisture contents, then the initial high rate infiltration exceeds the final rate by:

$$f_e(t) = \left[\frac{(c_{sat} - c_0)K_{sat} \cdot S_0}{\pi t} \right]^{1/2} \tag{12}$$

The only difference between the two solutions is the replacement of the numeric 2 in equation (11) by the numeric π in equation (12). The difference between these two solutions at 25% may appear large until it is recalled that the hydrologic parameters can vary by two orders of magnitude within a ten hectare field and do so in a manner that cannot be predicted by soil sampling. It is clear therefore that the problems raised by spatial variation are far more serious than those raised by the question of the precise choice of model for the investigation of the simplified problem of spatial homogeneity.

In the case of open channel flow, it can be shown that simplified forms of the linearized Saint-Venant equation can reproduce some of the properties of the complete linear equation. The general form of the linearized Saint-Venant equation for a prismatic channel of any friction law (Dooge et al 1986) is given by:

$$g \cdot \bar{y}_0 (1 - F_0^2) \frac{\partial^2 f}{\partial x^2} - 2u_0 \frac{\partial^2 f}{\partial x \partial t} - \frac{\partial^2 f}{\partial t^2} = \tag{13}$$

$$g \cdot A_0 (\frac{\partial S_f}{\partial Q} \cdot \frac{\partial f}{\partial t} - \frac{\partial S_f}{\partial A} \cdot \frac{\partial f}{\partial x})$$

where f(x,t) is the perturbation from the reference value of any hydraulic variable — the discharge rate Q(x,t), the area of flow A(x,t), etc. — and the derivatives of the friction slope with respect to the discharge and the area are evaluated at the reference conditions of flow. Since the equation is linear the solution is completely characterized by the linear channel response (i.e. the solution of equation (13) for a delta function upstream input) which can be convoluted with any input to obtain the corresponding output.

The shape of the linear channel response can be characterized by its moments or cumulants with respect to time. The effect of any simplification of equation 13 can therefore be studied by comparing the moments of the simplified equation with the moments of the complete linear solution. The most severe simplification is to neglect all the second-order derivatives in equation (13), thus equating the right-hand side of the equation to zero. The solution of this highly simplified model is

(14a) $$f(x,t) = f(t - \frac{x}{c_k})$$

where (14b) $$c_k = \left.\frac{dQ}{dA}\right|_0$$

is the kinematic wave speed of the given shape of channel and friction law. It can be shown (Dooge and Harley 1967, Dooge 1973, Dooge et al 1986) that the first moment of this kinematic wave solution is exactly equal to the first moment of the complete linear solution. If the first term on the left-hand side of equation (13) is retained then this simplified version of the complete equation is of convective-diffusive form. If instead of neglecting the second and third terms on the left-hand side completely, we use the kinematic wave solution to express them as ratios of the first term, we obtain an improved approximation. In the latter case, the first and second moments of the simplified equation are exact for all values of the Froude Number F and all of the moments are exact for a value of F = 0.

It is often argued that in this computer age simplified solutions of the type discussed above for various types of flow are no longer of any real value in hydrology. However, if we are to solve the scale problem and to succeed in parametrizing from a lower to a higher scale, this is most unlikely to be achieved without the use of simplified models at the lower scale. It should be recalled that in statistical mechanics the road to success involved the application of rationally based probability distributions to simplified models of behavior at the molecular scale.

3.2 Spatial variation in meso-scale hydrology

The comparison of alternative models on the basis of the solution for homogeneous one-dimensional modules as described in the last section provides some guidance for the synthesis of catchment response models but leaves unsolved the problem of the handling of spatial variation of non-linear elements in the hydrologic processes involved. The present section addresses directly the problem of spatial variation at the meso-scale and the approaches that can be made towards an understanding of its effects on the water balance of catchments and on the dynamics of run-off. In the discussion of subsurface flow the question of the relation of infiltration to other mechanisms of run-off production such as saturation overland flow and throughflow (Kirkby 1985, Troendle 1985) is deferred until a later section. In the case of surface flow the question is touched on briefly and dealt with again in the next section.

The study of the effect of spatial variability on the solution of the micro-scale equations governing the process of infiltration has been the subject of a number of studies over the past ten years. A listing of even the more important of these studies would be unduly lengthy in a Lecture such as this but ample references are available such as those listed in (a) the review paper by Dooge (1982) on "Parametrization of hydrologic processes" presented to the World Climate Programme Conference on Land Surface Processes in Atmospheric General Circulation Models held at Greenbelt, Maryland in January 1981; (b) in the review paper by Neuman (1982) on "Statistical characterization of aquifer heterogeneities: an overview" in the Geological Society of America Special Paper on Recent Trends in Hydrogeology; (c) in the review paper by Vauclin (1984) on "Infiltration in unsaturated soils" in the volume on Fundamentals of Transport Phenomena in Porous Media edited by Bear and Corapcioglu (1984); (d) in the papers presented to the workshop on "Soil spatial variability" and edited by Nielsen and Bouma (1985).

Some of these studies were concerned with the variation of such unsaturated soil parameters as the soil moisture and unsaturated hydraulic conductivity in the soil moisture zone and porosity and saturated conductivity in groundwater modules. Other studies were concerned with variation in the values of meso-scale parameters determined from field data such as sorptivity in the case of infiltration and transmissibility in the case of infiltration and transmissibility in the case of groundwater movement. From

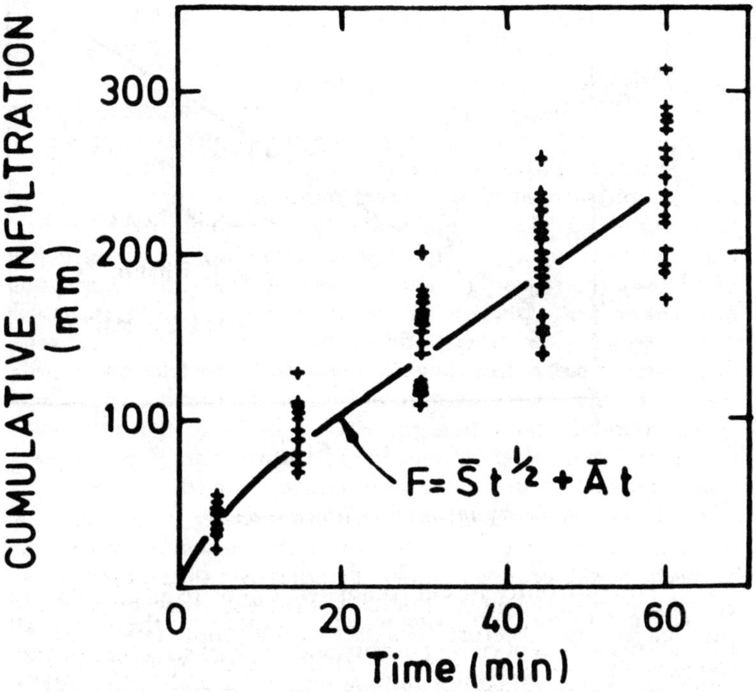

Figure 4. Infiltration at 17 points in hectare field.

the point of view of meso-scale hydrology the studies of field measurements and of parameters derived on the meso-scale are more useful and important than parameters based on soil samples or rock cores. The field studies of unsaturated flow were carried out on areas of an order of magnitude from ten hectares to one hundred hectares. In some studies the parameters were found to vary by two orders of magnitude and there appeared to be no definite link between this variation and the soil type or topography. Typical of these field studies are those by Sharma (1980) and Vauclin (1981). The approach to the study of spatial variability in meso-scale models can be by way of (a) deterministic analysis (e.g. Warrick and Amouzegar-Fard 1979) or by deterministic-stochastic analysis (e.g. Maller and Sharma 1981), (b) by numerical simulation in which the parameters are varied either in a fixed way (Peck et al. 1977, Sharma and Luxmoore 1979, Luxmoore and Sharma 1980) or by Monte Carlo method (e.g. Smith and Hebbert 1979), or by the direct use of stochastic differential equations (e.g. Bakr et al. 1978 and Gutjahr et al. 1979).

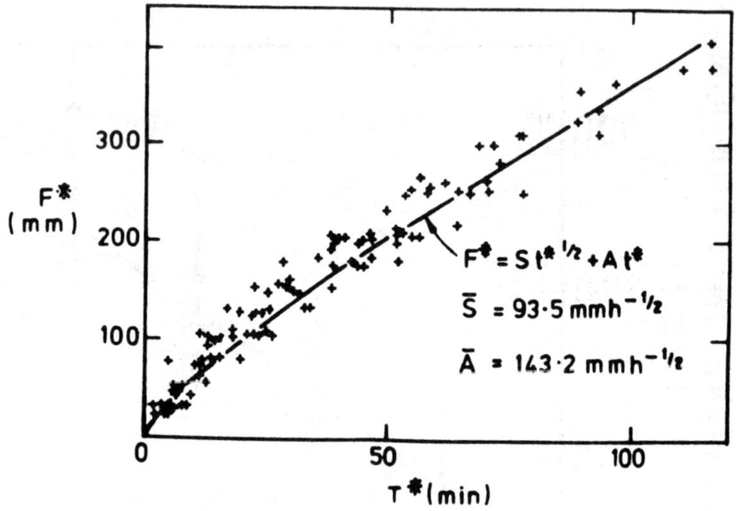

Figure 5. Soil similarity applied to infiltration data.

The variability in soil parameters can be reduced by use of the concept of soil similarity which assumes that the soil in all parts of the area is of equal porosity and internal structure, differing only in characteristic length scale (Miller and Miller 1956). Though there are the radical objections to the unrestricted use of such a scaling (Philip 1980), it has proved useful in reducing dramatically the scatter of empirical data both in the laboratory (Klute and Wilkinson 1958) and in the field (Warrick et al. 1977, Sharma et al. 1980, and Vauclin et al. 1981). Figure 4 shows infiltration at seventeen locations in a bare field of area one hectare. Figure 5 shows the dimensionless plot of the same data when data for the seventeen points are re-scaled on the basis of a characteristic length based on the value of the ultimate constant infiltration rate at each point.

One of the most difficult components to construct and operate in a conceptual model of catchment response is the soil moisture component (Fleming 1975, WMO 1975). The modelling of this component can vary from the use of a single uniform loss rate to a multi-layer model with complex rules for the input and the output from each of the layers. Variation in infiltration over the catchment area is frequently allowed for in such models by assuming a distribution of infiltration capacity with area. Many versions of this approach are capable of interpretation either in a deterministic or stochastic fashion. This is true of the Stanford model (Crawford and Linsley 1966), which uses a uniform distribution, and of models from the Soviet Union (Popov 1962), China (ECCHE 1977: 17–23), and Australia (Clark 1980) which use non-uniform distributions. Some more recent work (Moore and Clarke 1981, Moore 1985) is formulated on a strictly stochastic basis. Some of these models are based on infiltration capacity and thus are linked with Horton model of overland flow and others use a distribution of soil moisture storage and are linked with the concept of saturation overland flow. Most of the later models use a distribution function in the form of a power relationship rather than a uniform distribution.

Less work has been done on the problem of spatial variability in surface runoff. One example of deterministic modelling is the studies of Woolhiser on the effect of variation in slope between the elements of a cascade of overland flow segments on the kinematic wave solution (Kibler and Woolhiser 1970). An example of stochastic modelling of spatial variation is the work of Machado and O'Donnell (1982) who studied the effect on the response in the runoff from a plane surface when the input and the surface parameters (length, slope, roughness, and infiltration parameters) were selected from a normal distribution with a coefficient of variation of 10%.

3.3 *Conceptual modules in meso-scale hydrology*

As the scale of a hydrologic system increases, the lack of homogeneity makes it more and more difficult to apply models using physical equations which are based on the principles of continuum mechanics but with the intervention of two or more layers of parametrization. This difficulty led to the emergence in applied hydrology of the black-box approach in which input and output data sets were used to derive either a linear response or, by the use of Volterra series, a non-linear response. Intermediate between these two approaches is the assumption of plausible simple model whose parameters are calibrated from input-output data. Such models are often termed conceptual models and it is in this restricted sense that the term is used in the present section. Such models are usually formulated on the basis of a simple arrangement of a relatively small number of elements each of which is itself a simple representation of a physical relationship.

The section is restricted to deal with modules, i.e., with the representation of a single hydrologic process such as groundwater outflow, infiltration, overland flow, channel flow. The combination of such modules into a model of the total response of a subcatchment or a catchment is left until the next section. Such simple conceptual modules are of particular interest in the present context because their parameters can for the homogeneous case be related to physical parameters of the simplified solutions of the basic equations of physical hydrology. In the case of spatial variability this correspondence will break down but the relationship may still be useful in handling and interpreting the lumped parameters of the conceptual modules as derived from reliable measurements in the field.

It will be recalled from section 3.1 that for linearized one-dimensional unsteady groundwater flow, the predicted flow into a set of parallel drainage tiles or drainage ditches was independent of the method of linearization. No matter which of the two methods of linearization is used, we obtain for the impulse response of such a field drainage system the expression

$$h(t) = \frac{8}{\pi^2} \cdot \frac{i}{j} \sum_{n=1}^{\infty} \exp\left[-(2n-1)^2 \cdot \frac{t}{j}\right] \tag{15}$$

where j is the reservoir coefficient defined by Kraijenhoff van de Leur (1958) as:

$$j = \frac{1}{\pi^2} \cdot \frac{fS^2}{K\bar{h}} \tag{16}$$

where f is the drainable pore space, S is the spacing between drainage ditches or tile drains, K is the saturated hydrologic conductivity and $\bar{h}$ is the depth of groundwater used as the basis for linearization. The series represented by equation (15) is equivalent in all respects to a series of linear reservoirs of diminishing size, arranged in parallel. Except at very small values of elapsed time t, the first term will dominate the infinite series of equation (15). If the later terms in the series are neglected, then the impulse response becomes an exponential recession corresponding both to a widely used conceptual element (the single linear reservoir), and to the practice in applied hydrology of identifying a master recession curve for a given outflow by fitting a straight line to a semi-log plot.

Just as the outflow from groundwater can be simulated by a simple conceptual model which relates the groundwater outflow to the groundwater storage, so the simple analytical solutions for infiltration can be viewed as the behavior of simple conceptual elements in which the inflow into soil moisture storage is controlled by the amount of storage in the unsaturated zone, i.e., by the amount of water that has already infiltrated (Dooge, 1973). It is interesting to compare the two simple solutions discussed in section 3.1 above on the basis of contrasting assumptions in regard to the relationship of hydrologic diffusivity to moisture content by a comparison between the corresponding conceptual infiltration elements.

If it is assumed that the rate of infiltration excess over the final constant infiltration rate is inversely proportional to the total amount of water infiltrated, i.e.:

$$f_e = f(t) - f_0 = \frac{a_1}{F(t)} \tag{17}$$

it can readily be shown that this leads to the solutions given by equation (11) in 3.1 above where f_0 is the initial constant percolation rate corresponding to the constant initial moisture content c_0. Equation (17) gives an implicit solution for the cumulative infiltration F(t) in the form of the series:

$$\frac{f_0^2 t}{a_1} = \sum_{k=2}^{\infty} \frac{(-1)^k}{k} \cdot \left(\frac{f_0}{a_1} \cdot F\right)^k \tag{18}$$

which gives for the high rate infiltration at small values of the elapsed time:

$$f(t) = f_0 + \left(\frac{a_1}{2t}\right)^{1/2} \tag{19}$$

which is equivalent to equation (11) in section 3.1 above.

If, on the other hand, the rate of infiltration at the surface is taken as inversely proportional to the amount of excess high rate infiltration

$$f_e = \frac{a_1}{F(t) = f_0(t)} \tag{20}$$

this will lead to equation (19) directly.

The formulation of the infiltration problem in terms of such simple conceptual models requires the determination of the two parameters concerned either by estimation from micro-scale parameters and a particular micro-scale model, or else can be determined from measurements in the field at the meso-scale.

In the case of overland flow and open channel flow, the use of conceptual modules in applied hydrology started at a much earlier date. In the case of overland flow Horton (1933) extended the concept of a power relationship between steady overland outflow and equilibrium storage, which was supported by plot experiments on natural surfaces, to the more general concept of the existence of such a relationship during the unsteady flow phases of the rising hydrograph and the recession from equilibrium. This assumption corresponds identically to the representation of the overland flow surface as a single non-linear reservoir.

$$Q(t) = a[S(t)]^c \tag{21}$$

There is an exact relationship between the two parameters of the latter conceptual model and the hydraulic parameters of the steady state solution of the overland flow equation. The approach used by Horton for natural surfaces and solved by him for the case of $c = 2$ was extended to case of laminar flow over impermeable surfaces (where $c = 3$) through the analytical work of Keulegan (1944) and the experiments of Izzard (1944). In this case also the steady state solution can be used to relate the parameters of the hydraulic equation and the parameters of the conceptual module.

Conceptual modules were also used at an early date in solving problems of flood routing in open channels. Practicing hydrologists are aware of such conceptual modules for flood routing as the Muskingum method (McCarthy 1938), the Lag and Route method (Meyer 1941), and the Characteristic Reach method (Kalinin and Milyukov 1957). Each of these methods is a typical conceptual module. In the Muskingum method the storage in the channel reach is assumed to be a linear function of the upstream inflow and the downstream outflow, thus introducing two parameters to be determined either by linkage to the hydraulic parameters in the case of a prismatic channel or from a past flood event in the case of an irregular channel. The lag and route method assumes that the storage in the channel reach at any time is proportional to the outflow at some later time, the lag between the two times being constant. The characteristic reach method is based on the determination of the length of the channel for which storage can be assumed proportional to flow and the use of a cascade of such reaches to route the flood through the total length of channel.

These linear methods of hydrologic flood routing can be compared with one another and with the linearized Saint-Venant equation through the shape factors based on dimensionless moments already used in Figure 2 and Figure 3 in Section 2.3 above. If the dimensionless moments are based on the lag as in Figure 2, then the shape factors for the different routing methods plot as shown on Figure 6. The three conceptual modules pass through a

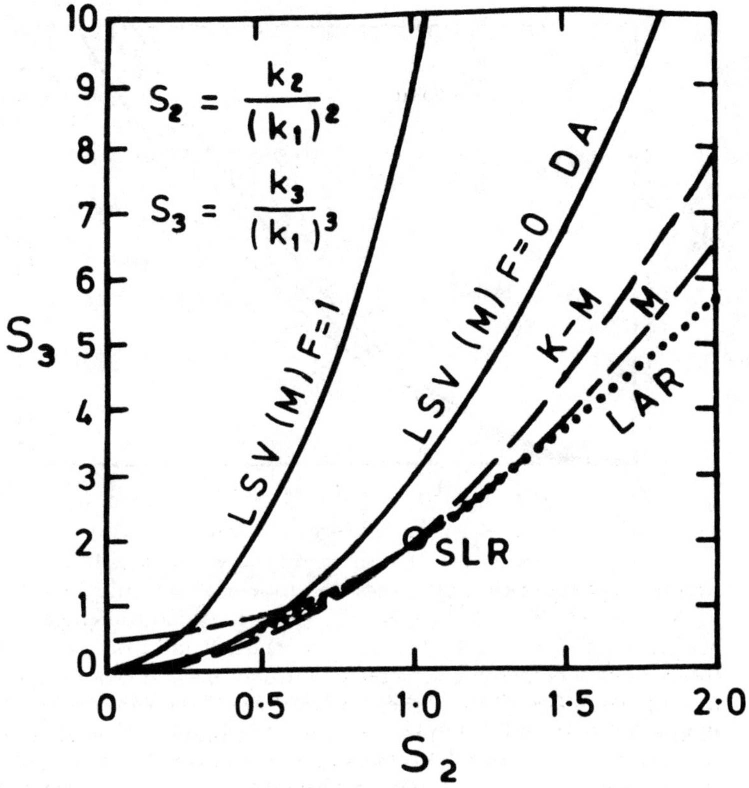

Figure 6. Shape factors for conceptual models.

common point corresponding to a single linear reservoir with no lag. It is clear from the figure that the match with the linearized Saint-Venant equation is not very good and that the Muskingum method matches very poorly for small values of s_2 which corresponds to long lengths of channel. If on the other hand the lag is allowed to become an extra free parameter, the methods can be compared by plotting dimensionless third and fourth moments based on the second moment as on Figure 3 in Section 2.3. The result is shown on Figure 7 in which the lag and route method reduces to a one-parameter model and plots only at the common point and the characteristic reach method gives a better match than the Muskingum method except for short reaches where $x < 0$ in the former and $n < 1$ in the latter method.

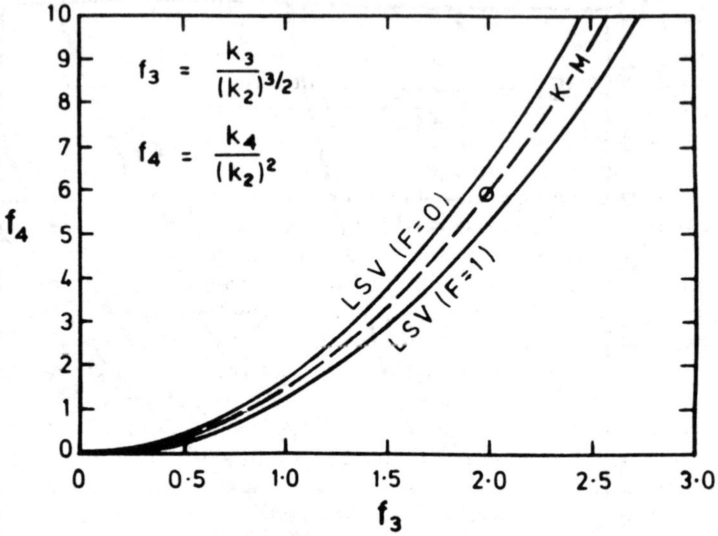

Figure 7. Corresponding shape factors based on variance.

3.4 Modules and systems

It has already been mentioned that as the scale of interest in hydrologic analysis becomes larger there is a tendency to shift from a meso-scale model based on the parametrization of the individual micro-scale equations to a meso-scale model of the whole system under examination. The approach based on the internal descriptions of physical hydrology suffers increasingly from the fact that we have in many problems a triple complexity arising from the geometrical complexity of the model required to reflect the variation in parameter values and the inter-connections in the system, the complexity of the equations themselves reflecting physical processes which are inherently non-linear, and the complexity of the inputs whose nature varies with space and time and season. A purely external description on the other hand suffers from the disadvantage that the inherent complexities make it extremely difficult to determine accurately the values of the parameters from the input and output data since the measurement is subject to error, the parameters are often highly correlated in some unknown fashion, and the spatial variability deprives the lumped parameters of physical significance so that they are unreliable for prediction purposes.

Even at the lower end of the meso-scale there may be a variety of mechanisms for the generation of catchment runoff operating within quite a small area. For many years the modelling of catchment response and the distinction between the rapid response and the slow response of the catchment was dominated by the concept of infiltration capacity and overland flow due to Horton (1933). Modules and models based on the alternative mechanisms of runoff generation by means of throughflow and saturation overland flow have been developed over the past 25 years. This development has been reviewed in the book on Hillslope Hydrology edited by Kirkby (1978) and in shorter contributions since then by Dunne (1982), Kirkby (1985), and Troendle (1985). From the point of view of hydrological prediction the importance of distinguishing between these mechanisms is not the nature of the hydrologic process involved but the fact that the lag times and hence the peak rate of flow are quite different for the different mechanisms for runoff generation. This is shown in Figure 8 due to Kirkby (1985) and based on data assembled by Dunne (1982). The occurrence of any one of these mechanisms could depend not only on highly variable local conditions within the catchment but also on the time of the year or on the nature of the rainstorm. To accommodate such a variety by means of an internal description based on hydrologic physics it would require a computer simulation of great size and expense.

The search for regularity and for acceptable simplifications is indeed a daunting task. However, there are indications that in the case of large catchments the super-position of a large number of elements, each of which is non-linear and spatially variable, may result in a total catchment response which will show a high degree of regularity and may be either linear or uniformly nonlinear. The fact that we are unable to prove that the aggregation of a very large number of non-isotropic water modules can safely be assumed to be an isotropic continuum does not deter us from basing all of the equations for the movement of water used in fluid mechanics and hydraulics on the latter assumption. It is worthwhile therefore to examine the question of whether such regularities are likely to emerge in the case of catchment response and whether there may not be a great deal more regularity in the total catchment response than in the behavior of the individual modules contributing to that total response. The bulk of this section is devoted to the discussion of an approach to this problem for the case of surface runoff.

The use of synthetic watershed models for simulation is used widely for hydrological prediction but is neglected as a tool towards hydrological understanding through numerical simula-

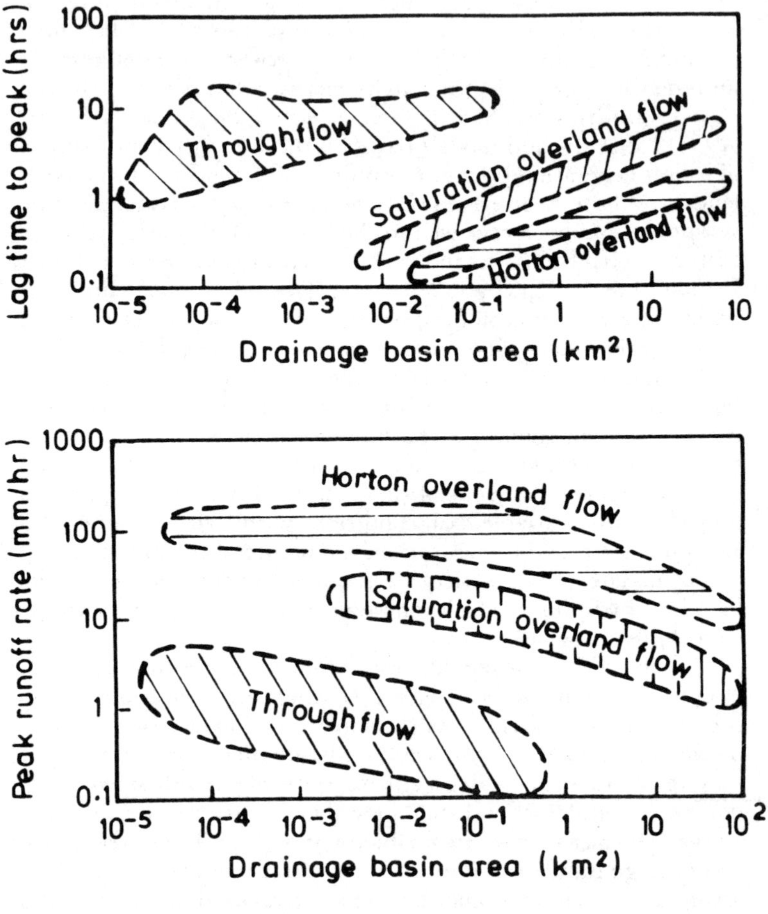

Figure 8. Mechanisms of surface runoff.

tion. Such numerical experimentation should be an aid to the generation of hypotheses and to selection of the more robust of them with a view to testing a small number of these hypotheses on real catchment data of a high quality. A model watershed was developed at the University of Minnesota with a view to studying the question of the non-linearity of surface runoff. (Machmeier 1966, Machmeier and Larson 1968). The Watershed model simulated 427 unit watersheds combined into a fourth-order catchment with an area of 21.35 square miles (55 square km). The lengths and slopes and cross-sections of the channels were based on the relation-

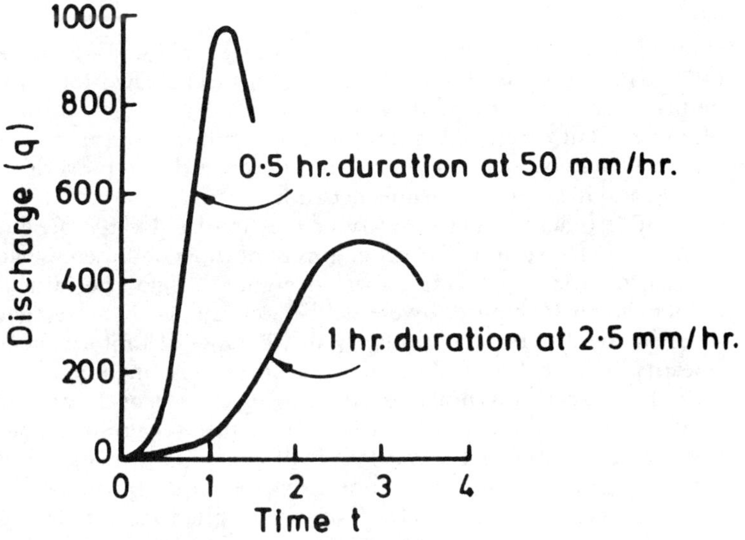

Figure 9. Hydrographs for constant volume.

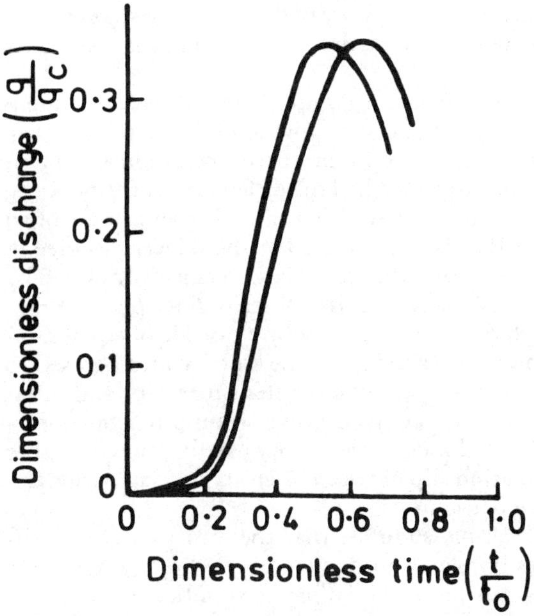

Figure 10. Dimensionless hydrographs.

ships of hydraulic geometry suggested by Leopold and Maddock (1953) and on some empirical relationships established for Minnesota catchments. Rainfall events were simulated by choosing a rate of rainfall excess and a duration, applying this input to the synthetic watershed and then routing the flow through overland storage and through the channel network.

The inherent non-linearity of the simulated catchment is naturally reflected in the hydrographs of outflow obtained as can be seen for the two events shown in Figure 9. The two rainfall events shown in Figure 9 were deliberately shown because they satisfied the criterion of similarity in the sense of uniform non-linearity (Dooge 1967). Since the Machmeier test included running the synthetic catchment to equilibrium for each of the rates of rainfall excess it is quite easy to test the hypothesis of uniform non-linearity by reducing the outflow hydrograph to dimensionless form by using the measured time to virtual equilibrium as the characteristic time by which the flow is multiplied and the elapsed time is divided. When this is done we obtain the dimensionless hydrographs shown in Figure 10 which clearly indicates a similarity in the form of the response function but a lag between the two events. As noted earlier, linear theory of flood routing suggests the existence of such a lag and also predicts no influence for the second moment of the response.

If the two hydrographs in Figure 9 are reduced to dimensionless form through the use of the time to peak the points from the two hydrographs define a common curve, thus indicating that in this complex simulated watershed the differences corresponding to third and fourth moments have been smoothed out or are of an order of magnitude that is not detected at the relevant degree of precision. This suggests that the complex watershed simulated by Machmeier with its 427 separate first order basins could be represented with a high degree of accuracy by a cascade of equal non-linear reservoirs and that results from one storm could be used to predict the runoff from a storm of a different intensity and duration, but obeying the similarity criterion of uniform non-linearity, by the use of the dimensionless response function together with a non-linear relationship between intensity and lag period derived from a number of events.

The above example suggests that there may at the meso-scale be regularities unrelated to micro-scale hydrology on which hydrologic models can be based. What is significant about the example based on the Machmeier and Larson data is that the behavior appears to be uniformly non-linear even though the degree of non-linearity is different in the modules of overland flow and the modules of channel flow. If the hypothesis of uniform non-linearity for catchments is valid for catchments of a given size then the derivation of the Volterra kernels of the non-linear response is greatly simplified as shown by Napiorkowski (1985).

There are indications that searches for regularity at the upper end of the meso-scale may be worthwhile for problems other than that of isolated flood events discussed in this section. Eagleson (1978) applied a stochastic dynamic approach to the problem of the longterm water balance using simplified representations of micro-scale hydrologic processes together with assumed probability distributions to derive the relationship between the key ratio of actual to potential longterm evapotranspiration and soil and vegetation parameters. The application of this approach to the problem of flood frequency is discussed in the present section. Klemes (1978, 1986) has discussed the importance of giving a hydrologic content to the stochastic models used in meso-scale hydrology.

The search for regularities at the upper end of meso-scale may well yield to a combined stochastic-dynamic approach (Eagleson 1972, Klemes 1978, Zhu 1985) analogous to the pioneering work in statistical mechanics in the last century. Eagleson (1972) combined two simple probabilistic assumptions (one relating to rainfall and the other to rainfall-runoff transformation) with the Wooding (1966) model of runoff routing involving the kinematic wave solution for rectangular areas of plane overland flow draining to a stream of constant cross-section and slope. By combining these simple probabilistic and deterministic assumptions, Eagleson (1972) derived a distribution function for flood peaks involving four parameters which were contained in the basic assumptions and thus can be given a physical meaning.

Versions of this combined probabilistic deterministic approach were later used (Hebson and Wood 1982, Diaz-Granados et al. 1984) which based the rainfall-runoff transformation on the geomorphic unit hydrograph which will be discussed in the next section. The approach reveals some promise but is still at a relatively early stage of development (Bras et al. 1985).

4. MACRO-SCALE MODELS IN HYDROLOGY

4.1 The Drainage Network

The search for hydrologic laws is a search for simple relationships amid complex phenomena. As an alternative to building up from the micro-scale through the meso-scale by parametrization of the laws based on continuum mechanics, we can attempt to seek laws directly on the macro-scale itself. As in any such endeavor, the correct approach (Polya 1957) is to solve the problem in the simplest possible form and then to proceed to the more complex problem. Accordingly, we find in the approaches to macroscale hydrology, which are still at an early stage, a concentration on simpler problems of long term relationships in which we use a large time-scale as well as a large space-scale. A notable attempt at finding a source of regularity and simplicity at the catchment scale has been based on the attempts to develop laws resulting from the geomorphological equilibrium of the drainage channel network. Such an approach was pioneered by Horton (1945a, 1945b) and Rzhanitsyn (1960) and useful reviews of the approach have been published by Smart (1972) and Zavoianu (1985).

The drainage network of a catchment links the upstream sources where the runoff first appears in the definite form of a channel flow to the outlet of the main stream. The first step towards developing a theory of drainage network patterns, whether empirical or theoretical, is to classify the channels in the network on the basis of what is known as order. There are a number of alternative ways of defining the order of streams in a channel network. In the later work on the subject the channel reach between two junctions is termed a link and it is the links that are given a definite order. The classification most often used is that due to Strahler (1952). In the Strahler method of classification channels originating at a source are defined to be first order streams; when two streams having the same order meet at a junction the channel downstream of the junction is taken as being one order higher than the order of the two upstream links; however, when two streams of different order meet at a junction the downstream link retains as its order the higher of the two orders of the streams meeting at the junction. It is assumed that multiple junctions do not occur.

Horton (1945a, 1945b) proposed empirical laws to the effect that both the number of streams of successive orders and the mean length of streams of successive orders could be approximately represented as a geometric series. The common ratio of the geometric series of stream numbers was called the bifurcation ratio and was found by examination of a large number of catchments to take on a value between 3 and 5 in nearly all cases. The common ratio between the mean stream length of successive orders was found by Horton to be between 1.5 and 3.5. Horton also suggested that the stream slopes would follow a geometric series but did not provide empirical data in support of this proposition. Schumm (1956) proposed a further geometric law for the drainage areas of streams of successive orders.

A significant development in the study of drainage networks was the work of Shreve (1966) who replaced Horton's empirical deterministic approach by a stochastic approach based on the concept of a topologically random population of channel networks. Shreve (1966) defined such a topologically random population as one in which all topologically distinct networks for a given number of sources are equally likely. He developed formulae for the relative probability of different sets of stream numbers in such a network population and showed that the most probable network for any given magnitude (i.e., given number of upstream sources) conformed closely to Horton's Law of stream numbers. Later Gupta and Waymire (1983) showed analytically that the expected value of the number of streams of successive order approached a geometric series with a stream ratio of 4 as the number of sources (i.e., the magnitude of the drainage network) became very large. It has already been remarked that in actual catchments the bifurcation ratio has been found to vary between 3 and 5 except in cases where there is a very marked local geological control which inhibits the development of a natural drainage network.

The next series of developments in this approach were connected with the attempt to link hydrologic variables to the topological parameters of the drainage network. Rzhanitsyn (1960) developed some relationships between stream order and different measures of mean and maximum discharge. Boyd (1978) studied the relationship between mean lag time and the topological network parameters for a fourth-order catchment area of 40 km^2. The mean lag times were also found to form a geometric series.

The linking of mean discharge and catchment order is illustrated in Figure 11 due to Zavoianu (1985). Figure 11(a) shows the standard plot of the number of streams against stream order for the River Doftana at its confluence with the Prahova and for the River Varbilau at its confluence with the Teleajen. Figure 11(b) shows a plot against stream order of the product of the order by the sum of the mean discharges for all catchments of that order. In this case a geometrical series is also obtained with a common ratio very close to unity. Figure 11(c) shows the plotting against stream order of the average mean discharge of streams of that particular order. The particular case of the River Varbilau is interesting because of the break of slope shown in Figure 11(b), i.e. in the case of the summed mean discharges (Zavoianu 1985). This reflects the occurrence in the Varbilau catchment of two differing sub-catchments: an upper sub-catchment where precipitation exceeds evaporation and a lower sub-catchment where evaporation slightly exceeds precipitation. It is interesting to note that the break occurs in the conventional plot of stream order against number of streams of that order in

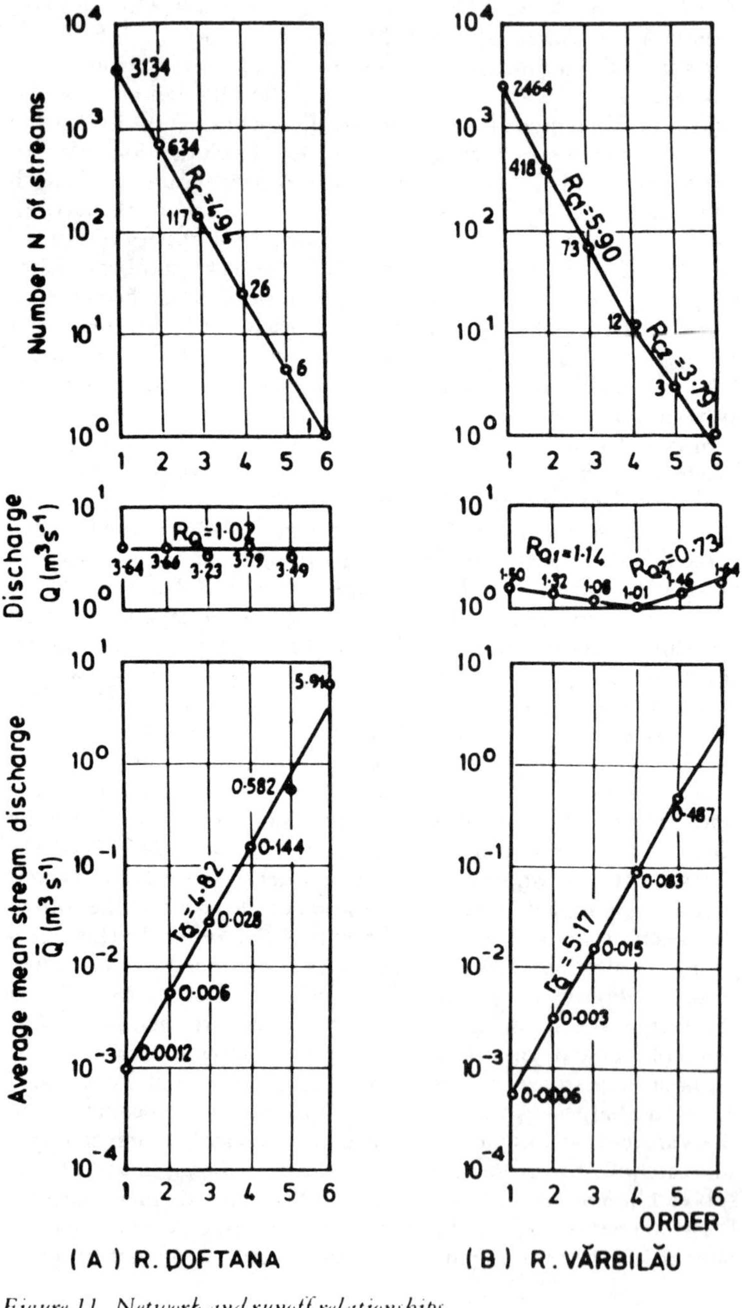

Figure 11. Network and runoff relationships.

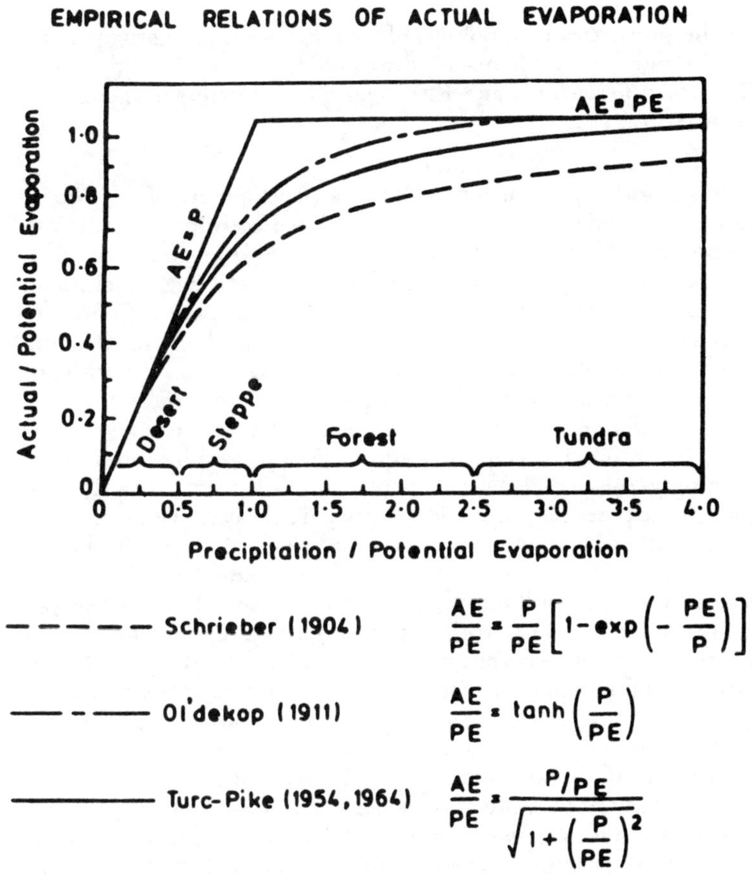

Figure 12. Empirical relations for actual evaporation.

Figure 11(a) but does not occur in the plot of stream order against mean discharge of the same order.

Rodriguez-Iturbe and Valdez (1979) studied the properties of the geomorphic unit hydrograph, i.e. the response of a catchment which conforms to the Horton-Shreve laws of drainage composition. Typical of their results was the derivation of a relationship between the product of the peak flow and time to peak to the ratio of the bifurcation ratio R_B and the area ratio R_A. This approach was generalized somewhat by Gupta, Waymire and Wang (1980). Further work linking the instantaneous unit hydrograph to the properties of topologically random channel networks was carried out by Kirshen and Bras (1983) and by Karlinger and Troutman (Troutman and Karlinger 1985, Karlinger and Troutman 1985).

4.2 *Climate, Soil and Vegetation*

Another general approach to the development of hydrologic laws on the macro-scale seeks to establish equilibrium relationships for the elements of the long-term water balance as a preliminary to the study of dynamic relationships between the climate, the soil, the vegetation and the hydrology of the area. Central to this general approach is the key hydrological problem of reducing potential evapotranspiration as determined by micro-climatic conditions to actual evapotranspiration from the land surface.

Under very wet conditions there will never be any shortage of water in order to supply potential evapotranspiration and accordingly the actual evapotranspiration will asymptotically approach the potential value the more humid the conditions become. On the other hand under very dry conditions the excess of the potential evapotranspiration over actual precipitation will be such that all the water will be evaporated and we have the actual evapotranspiration equal to precipitation. Figure 12 shows a number of relationships between the ratio of actual to potential evapotranspiration as a function of the ratio of precipitation to potential evapotranspiration as determined by the local climate. It can be seen that in each case the estimate of actual evapotranspiration asymptotically approaches the precipitation for very dry conditions and asymptotically approaches the value of potential evapotranspiration for very wet conditions. The ratio of precipitation to potential evaporation is the reciprocal of the aridity index due to Budyko (1971) and the classification indicated on the ordinate is due to him.

The upper curve in figure 12 is an exponential relationship due to Schreiber (1904).

$$AE = P[1 - \exp(-\frac{PE}{P})] \tag{22}$$

The lower relationship is a hyperbolic tangent due to Ol'dekop (1911).

$$AE = PE[\tanh(\frac{P}{PE})] \tag{23}$$

Budyko (1948, 1961, 1971) analyzed the long-term water balance data for over a thousand catchments in the USSR and found that the values for the individual catchments plotted somewhere in between these two relationships. Consequently, Budyko (1971) proposed that the geometric mean of the estimates given by these two limiting equations should be used in order to estimate the actual evapotranspiration and indicated that the use of such an equation would represent the data within ±10%. Turc (1954, 1955) proposed an alternative equation based on data from a large number of catchments in Africa. Turc's formula was later slightly modified by Pike (1964) and the modified form is shown as the central full curve in figure 12. The relationship indicated by the Turc-Pike equation and that indicated by the Budyko equation are virtually identical and in the opinion of the author the former relationship given by:

$$AE = \frac{P}{[1 + (P/PE)^2]^{1/2}} \tag{24}$$

which is simpler in form than the Budyko relationship could be taken as an empirical starting point for the study of such long-term relationships.

Eagleson (1978) used a stochastic-dynamic approach to this problem of the long-term water balance. He assumed representative probability density functions for such climatic variables as the interval between storms, the duration of storms, the intensity of rainfall in the storms and the potential evapotranspiration. Combining these assumed probability density functions with simplified assumptions in regard to the hydrological processes involved in the conversion of storm rainfall to runoff, Eagleson (1978) derived the probability density function of the actual infiltration during storms and the probability density function of the actual evaporation between storms. The average volume of infiltration and the average volume of evaporation can then be found from the average number of storms and used to construct the long-term water balance.

Using this approach, Eagleson (1978) was able to relate the key ratio of actual to potential evapotranspiration to 5 parameters; three of these parameters related to the properties of soil and two of them to the properties of the vegetation. The resulting relationship has the same general form as the Turc-Pike relationship shown on figure 12 but has the additional advantage that the effect of changes in the soil parameters and the vegetation parameters and in the statistical assumptions about climatic can be systematically studied.

In a further development of this approach to the long-term water balance, Eagleson (1982) suggested the existence of equilibrium relationships between the long-term actual evapotranspiration and the state of the vegetation canopy. Eagleson introduced two hypotheses, one relating to the situation where the vegetation is water-limited and the second where the vegetation is energy-limited. Eagleson (1982) suggested that under conditions of water-limitation, a system of vegetation would for the given climate and soil moisture conditions produce the particular canopy density which reduced moisture stress at the roots to a minimum. For the case where vegetative activities are limited by energy rather than by water, Eagleson (1982) suggested that the vegetative system would tend to maximize the biomass for the given amount of energy. By applying these two hypotheses to his 1978 stochastic-dynamic model, Eagleson (1982) derived the equilibrium relationship defining the limiting curves relating the ratio of actual to potential evapotranspiration (which is species-dependent) to the density of the vegetative canopy. Preliminary comparison of data for a few catchments in humid and semi-arid regions tends to confirm the derived limiting relationships as reasonable (Eagleson and Tellers 1982). In a further development (Eagleson and Segarra 1986) these hypotheses of ecological optimality were applied to the effect of climate change on the annual water balance of Savanna vegetation and again the results are encouraging.

The need for hydrologic techniques applicable at the large scale has been highlighted by the recent worldwide interest in the problem of climate variation and change and in the resulting impacts, whether physical, biological, social or economic (WMO 1980, WMO 1986). The conference of experts convened at Villach by WMO, UNEP and ICSU in 1985 expressed the consensus that as a result of the increase of CO_2 and other greenhouse gases in the atmosphere the annual mean runoff may increase in high latitudes, but summer dryness may become more frequent at middle latitudes in the northern hemisphere, and that potential evapotranspiration will probably increase throughout the tropics whereas in moist tropical regions convective rainfall could also increase (WMO 1986).

In a paper presented to the World Climate Conference, Schaake and Kaczmarek discussed the general problem of the relationship of climate variation and change to water resources development. Some of the work done on the problem since then has recently been reviewed by Klemes (1985). It can readily be shown that climatic changes, whether due to an increase in atmospheric carbon dioxide or otherwise are subject to hydrological magnification but as yet there are no agreed hydrological techniques for estimating the degree of this amplification (Dooge 1986b). The effect of changes in long-term runoff could have a most serious effect on the performance of reservoirs systems now being planned and involving very large amounts of capital expenditure. The resolution of the problems involves a real challenge to the hydrologic community. The scheduling by IAHS of a symposium on the "Influence of climate changes and climate variability on the hydrological regime and water resources" for the Vancouver General Assembly for 1987 is indeed timely.

4.3 Global hydrology

Since the fourth Kisiel Memorial Lecture delivered a year ago (Eagleson 1986) dealt with the topic "The Emergence of Global-Scale Hydrology," there is little need to deal with this particular topic in this the fifth Memorial Lecture except for the sake of completeness. Any account of scale problems in hydrology would be incomplete without a discussion, however short, of some aspects of the global or planetary scale.

Every textbook on hydrology gives a table of the distribution of water on our planet among various forms of storage and in nearly every case the impression is given that this distribution is accurately known. In fact we do not know with any certainty some of the items included in these tables and most of them are presented as closed balances in which the effect of changes in the mean sea-level or changes in the water content of the ice-caps are ignored (Szesztay 1970). Even when these variations are taken into account there are still unexplained discrepancies (Meier 1983) and a substantial problem exists in this connection which can only be solved by the cooperation of scientists from a number of branches of geophysics.

Another source of great uncertainty is the amount of inactive groundwater. Some tables do not even distinguish between active and inactive groundwater. Table C due to Kalinin (1968) does make this distinction and defines the total groundwater as being the groundwater down to a depth of 5 km below the surface. This table shows the great variety in time-scales which are operative at the global scale. Thus the residence time of water in the frozen state in the cryosphere is estimated at over 15,000 years, the residence time in the ocean at 3,000 years and the residence time in the active ground-water as 300 years. In contrast to these long time scales, the residence time of water stored in the atmosphere, in rivers and in the biosphere are all of the order of 10 days and the storage of water in the soil on which our food supply depends is only about 300 days.

TABLE C

STORAGE AND REPLENISHMENT OF GLOBAL WATER

TYPE OF STORAGE	AMOUNT (10^6 km^3)	FLUX (10^3 km^3)	MEAN RESIDENCE TIME
TOTAL	1460	E = 520	1800 YEARS
OCEANS	1370	E = 449	3100 YEARS
INACTIVE GROUNDWATER	56		
FROZEN WATER	29	R = 1.8	1600 YEARS
ACTIVE GROUNDWATER	4	R = 13	300 YEARS
SOIL WATER	65×10^{-3}	E + R = 85	280 DAYS
ATMOSPHERE	14×10^{-3}	P = 520	9 DAYS
RIVERS	1.2×10^{-3}	R = 36	12 DAYS
BIOLOGICAL WATER	0.7×10^{-3}		7 DAYS

Once we attempt to break down the water balance to a continental scale or a regional scale, we enter an area in which estimates can be made (Baumgartner and Reichel 1975, Korzun et al 1974) but of which our understanding is very limited. Kalinin (1968) has pointed out the existence of apparent teleconnections in the runoff of large catchments which can only be adequately explained when we have a thorough understanding of the general circulation of some large-scale features in the earth's atmosphere. Table D due to Kalinin (1968) shows the correlation of the annual flows of the Yellow River in China, the Mississippi in the United States and the Volga in the Soviet Union. We expect the annual runoff from neighboring catchments to be highly correlated but for the correlation to decrease relatively rapidly with distance. The existence of the co-efficient of correlation of 0.48 between the annual flows on the Mississippi and the Volga which are on opposite sides of the planet can scarcely be dismissed as being due to chance and indicates existence of large-scale phenomena which we have not yet begun to understand. The further circumstance that the Huang Ho which is widely separated from the other two appears to be quite uncorrelated with the Mississippi but highly negatively correlated with the Volga only adds to the mystery.

Every hydrologist is aware of the immense difficulties created by the shortness of the instrumental hydrological record. It

TABLE D

CORRELATION OF ANNUAL FLOWS

	HUANG HO	MISSISSIPPI	VOLGA
HUANG HO (105°E)	–	– 0.08	– 0.55
MISSISSIPPI (105°W)	– 0.08	–	0.48
VOLGA (40°E)	– 0.55	0.48	–

may well be that laws of a catchment hydrology at a macro-scale can only be developed by extending the record back to pre-historic times through the techniques of modern palaeo-hydrology and the careful analysis of the results (Schumm 1967, Kochel and Baker 1982, Gregory 1983, Baker 1985b). If we are to develop hydrology on a planetary scale then it will be necessary to be able to make deductions about the hydrology in past times on the inner planets of the solar system. The question of the channels on Mars has long been a subject of speculation and controversy and indeed is still a matter of controversy today (Baker 1982, Mars Channel Working Group 1983, Baker 1985a). Even if the more immediate problems of meso-scale hydrology raised during this discussion are resolved within the next decade, there will be no shortage of problems to be solved at the macro-scale.

No matter which of the nine spatial scales of Table A we are concerned with and whether we are attempting to parametrize from a lower scale or disaggregating from a higher scale, certain qualities and attitudes displayed by Chester Kisiel who is commemorated in this Lecture — an open mind, a capacity for hard work, a familiarity with the work done not only in hydrology but in neighboring disciplines, and an ability to co-operate with others — are all needed. With these qualities, Chester Kisiel and his contemporaries brought about a revolution in hydrologic methods. Enough remains to be done to allow each of several succeeding generations of hydrologists to achieve their own revolution in their own time.

References Cited

Aranon, R. H., "Statistical approach to flow through porous media," *Phys. Fluids*, 9:1721, 1966.

Aravin, V. I. and S. N. Numerov, *Teoriya dvizheniya jhidokstei i gazov v nedeformiruemoi porostai srede*, G.I.T. — T.L. Moscow, 1953. Trans. into English by A. Moscona as *Theory of fluid flow in underformable porous media*, Israel Program for Scientific Translations, Jerusalem, 1965.

Bachmat, Y., "Spatial macropization of processes in heterogeneous systems," *Israel J. of Tech.*, 10(5):391–403, 1972.

Baker, V. R., *The channels of Mars*. University of Texas Press, Austin, Texas, 1982.

____, "Models of fluvial activity on Mars," *Models in geomorphology*, (ed. M. Woldenburg), pp. 287–312, Allen and Unwin. London, 1985a.

____, "Palaeoflood hydrology of extraordinary flood events," Proceedings of U.S.-China bilateral symposium on the analysis of extraordinary flood events, *Journal of Hydrology*, October, 1985.

Bakr, A. A., L. W. Gelhar, A. L. Gutjahr, and J. R. McMillan, "Stochastic analysis of spatial variability in sub-surface flows. Part I: comparison of one- and three-dimensional flows," *Water Resources Research*, 14(2):263–271, 1978.

Baumgartner, A. and E. Reichel, *The World Water Balance*, Elsevier, Amsterdam, 1975.

Bear, J., *Dynamics of Fluids in Porous Media*, Elsevier, New York, 1972.

Bear, J. and Bachmat, "Transport phenomena in porous media — basic equations," *Fundamentals of Transport Phenomena in Porous Media*, (Ed. J. Bear and M. Y. Corapcioglu), pp. 3–6, Martinus Nijhoff, Dordrecht, 1984.

Benedict, F. G., *Vital Energetics: a Study in Comparative Basal Metabolism*, Carnegie Institute of Washington, 503, Washington D.C., 215 p, 1938.

Bennion, D. W. and J. C. Griffiths, "A stochastic model for predicting variations in reservoir rock properties," Trans, *AIME*, No. 237(2):9–16, 1966.

Beran, M. J., *Statistical Continuum Theory*, Interscience, New York, 1968.

Boyd, M. J., "A storage-routing model relating drainage basis hydrology and geomorphology," *Water Resources Research*, Vol. 14(5):921–928, 1978.

Bras, R. L., M. S. Moughamlan, and D. B. McLaughlin, "Estimation of flood frequency: a comparison of physically based procedures," Proceedings of U.S.-China bilateral symposium on the analysis of extraordinary flood events, *Journal of Hydrology*, 1985.

Brush, S. G., *Statistical Physics and the Atomic Theory of Matter from Boyle and Newton to Landau and Onsager*, Princeton University Press, 1983.

Budyko, M. I., *Evaporation under natural conditions*, Gimiz, Leningrad, 1948. IPST, Jerusalem, 1963.

Budyko, M. I. and I. L. Zubenok, *Opredelenie ispareniya s poverkhosti sushi*, (The determination of evaporation from the land surface) Izv. Ak. Nauk. SSSR. Ser. Geog. 6:3–17, 1961.

Budyko, M. I., *Klimat i Zhizn*, Geidrometeor. Izdat, Leningrad, 1971. Trans. by D. H. Miller as *Climate and Life*, Academic Press, New York, 1974.

Charnyi, I. A., "Strogeo dokazatel'stvo formuly Dyupyui dlya beznapomoi filtratsii s promezhutkom vysachivaniya," (A rigorous proof of the Dupuit formula for unconfined seepage with a seepage face), Doklady, AN SSSR, Vol. 79, No. 6, 1951.

Clark, R. D. S., "Rainfall stormflow analysis to investigate spatial and temporal variability of excess rainfall generations," *Journal of Hydrology* 47:91–10, 1980.

Crawford, N. H. and R. K. Linsley, "Digital simulation in hydrology. Stanford Watershed Model IV," Department of Civil Engineering, Stanford University, Technical Report No. 39, 1966.

Cullingford, R. A., D. A. Davidson, and J. Lewin, *Time Scales in Geomorphology*, Wiley, 1980.

Dagan, E. and E. Bressler, "Solute dispersion in unsaturated heterogeneous soil at field scale: I theory," *Soil Sci Soc. Am. Proc.*, 1979.

Darcy, H., *Les Fontaines Publiques de la Ville de Dijon*, Victor Dalmont, Paris, 1856.

Davis, S. N., "Porosity and permeability of natural materials," *Flow through Porous Media*, (ed. R. J. M. de Wiest), pp. 54–89, Academic Press, New York, 1969.

Diaz Granados, M. A., J. B. Valdes, and R. L. Bras, "A physically based flood frequency distribution" *Water Resources Research*, 20(7): 295–1002, 1984.

Dooge, J. C. I., "A new approach to non-linear problems in surface water hydrology," Proceedings of IAHS General Assembly of Berne, *Int. Assn. of Sci. Hydrology Publ.*, 76, pp. 109–413, 1967.

———, "Linear Theory of Hydrologic Systems," Technical Bulletin No. 1468, U.S. Agricultural Research Service, Washington, 327 pp., 1973.

———, Flood Routing in Channels, Postgraduate Lectures, Department of Civil Engineering, University College Dublin, 1980.

———, "The parametrization of hydrologic processes," *Land Surface Processes in Atmospheric General Circulation Models,* (ed. P. S. Eagleson), pp. 243–288, Cambridge University Press, 1982.

———, "On the study of water," *Hydrological Sciences Journal,* 28(1): 23–48, 1983.

———, "Looking for Hydrologic Laws," *Water Resources Research*, 22(9):465–585, 1986a.

———, "Climate change and water resources" Lecture to W.M.O. Executive Council, *W.M.O.*, Geneva, 1986b.

Dooge, J. C. I., and B. M. Harley, "Linear routing in uniform open channels," *Proc. International Hydrology Symposium,* Vol. 1, 57–63, Ft. Collins, Colorado, 1967.

Dooge, J. C. I., J. J. Napiorkowski, and W. G. Strupczewski, "The linear downstream response of a generalized uniform channel," *Acta Geophysica Polonica,* 1986.

Dullien, F. A. L., *Porous Media Fluid Transport and Pore Structure,* Academic Press, New York, 1979.

Dunne, T., "Models of runoff processes and their significance," *Studies in geophysics: Scientific Basis of Water Resources Management*, (ed. J. R. Wallis), 19–30, National Academy of Sciences, Washington, 1982.

Dybbs, A. and R. V. Edwards, "A new look at porous media fluid mechanics — Darcy to turbulent," *Fundamentals of transport Phenomena in Porous Media,* (ed. J. Bear and M. Y. Corapcioglu), pp. 199–155, Martinus Nijhoff, Dordrecht, 1984

Eagleson, P. S., "Dynamics of flood frequency," *Water Resources Research,* 8(4):878–898, 1972.

———, "Climate, soil and vegetation," *Water Resources Research,* 14(5): 705–776, 1978.

———, "Ecological optimality in water-limited natural soil-vegetation systems. I: theory and hypothesis," *Water Resources Research* 19(2):325–340, 1982.

———, "The Emergence of Global-Scale Hydrology, Fourth Chester C. Kisiel Memorial Lecture, Department of Hydrology and Water Resources, University of Arizona, Tucson, 1986.

Eagleson, P. S. and R. I. Segarra, "Water-limited equilibrium of savanna vegetations systems," *Water Resources Research*, 21(10):1483–1493, 1985.

Eagleson, P. S. and R. R. Tellers, "Ecological optimality in water-limited natural soil-vegetation systems. 2. Tests and applications," *Water Resources Research*, 18(2):341–354, 1982.

Eames, C. and R. Eames, Powers of Ten, 16 mm film (9½ minutes), Pyramid Films, P.O. Box 1048, Santa Monica, California, 1977.

ECCHE, *Flood Forecasting for Humid Regions of China*, East China College of Hydraulic Engineering, Nanjing, China, 1977.

Eddington, A. S., *New Pathways in Science*, Cambridge University Press, 1935.

Einstein, A., "Ueber die von der molekular-kinetischen Theorie der Warme geforderte Bewegung von in ruhenden Flussigkeiten suspendierten Teilchen," *Annalen der Physik Ser.* 4. 17:549–560, (1905). Trans. into English as "On the movement of small particles suspended in a stationary liquid demanded by the molecular kinetic theory of heat," reprint, *Investigations on the Theory of Brownian Movement*, Dover Publications, New York, 1926.

Farengolts, Z. D. and M. N. Kilyada, "Opyt primeneniya metodov matematicheskoy statistiki dlya izutcheniya zakonov raspredeleniya gidrogeologicheskikh parametrov," *Trudy Vsegingeo*, 17:76–112, 1969.

Fiering, M., The Real Benefits from Synthetic Flows, Second Chester C. Kisiel Memorial Lecture, Department of Hydrology and Water Resources, University of Arizona, Tucson, 1984.

Finkl, C. W., "Soil Classification," *Benchmark Papers in Soil Science*, Hutchinson Ross, U.S.A., 391 pp, 1982.

Fleming, G., *Computer Simulation Techniques in Hydrology*, Elsevier, New York, 333 pp, 1975.

Freeze, R. A., "A stochastic conceptual model analysis of one-dimensional groundwater flow in a non-uniform homogeneous media," *Water Resources Research*, 11(5):725–741, 1975.

Gelhar, L. W., "Effects of hydraulic conductivity variations on groundwater flows," *Proceedings of Second IAHR Symposium on Stochastic Hydraulics*, (Lund, Sweden), International Association for Hydraulic Research, 1976.

Gray, W. G. and K. O'Neill, "On the general equations for flow in porous media and their reduction to Darcy's law," *Water Resources Research*, 12(2):148–154, 1976.

Green, W. H. and G. A. Ampt, "Studies on solid physics: the flow of air and water through soils," *Journ Agr. Res.*, 4:1–24, 1911.

Gregory, K. J., *Background to Palaeohydrology*, Wiley, 1983.

Gupta, V. K., E. Waymire, and C. T. Wang, "A representation of an instantaneous unit hydrograph from geomorphology," *Water Resources Research*, 6(5):855–862, 1980.

Gupta, V. K. and E. Waymire, "On the formulation of an analytical approach to hydrologic response and similarity at the basis scale," *J. Hydrology*, 65:95–123, 1983.

Gutjahr, A. L., L. W. Gelhar, A. A. Bakr, and J. R. MacMillan, "Stochastic analysis of spatial variability in subsurface flows. Part II: evaluation and application," *Water Resources Research*, 14(5):953–960, 1978.

Hebson, C. and E. Wood, "A derived flood frequency distribution," *Water Resources Research*, 18(5):1509–1518, 1982.

Hollis, C., *The American Heresy*, Sheed and Ward, London, 1927.

Horton, R.E., "The role of infiltration in the hydrological cycle," *Trans. Am. Geophy. Union.*, 14:446–460, 1933.

——, "The interpretation and application of runoff plot experiments with reference to erosion problems," *Soil Sc. Soc. Am. Proc.* 3:340–349, 1945a.

——, "Erosional development of streams and their drainage basins: hydrophysical approach to quantitative geomorphology," *Geol. Soc. Amer. Bull.*, 56:275–370, 1945b.

Hubbert, M. K., "The theory of ground-water motion," *J. Geol.*, 48:785–944, (1940). Reprint, *The Theory of Ground-Water Motion and Related Papers*, M. K. Hubbert, Hafner, New York, 1969.

Izzard, C. F., "The surface profile of overland flow," *Trans. Am. Geophys. Union*, 25(6):959–968, 1944.

Jetel, J., "Complement regional del'information sur les parametres petrophysiques en vue de l'elaboration des modeles de systemes aquifere," *Proc. of Montpelier Meeting*, International Association of Hydrogeologists, 1974.

Kachanoski, R. G., "Processes in soils: from pedon to landscape," SCOPE Workshop on Spatial and Temporal Variability of Biospheric and Geospheric Processes, St. Petersburg, Florida, Oct.-Nov. 1985.

Kalinin, G. P., *Problemy Global'noi Gidrologii*, Geidromet. Iszdat, Leningrad, 1968. Trans. into English, (Global Hydrology), Israel Programme for Scientific Translations, Jerusalem, 1971.

Kalinin, G. P. and P. I. Milyukov, "O Raschete Neustanovivshegosya Dvigheniya Vody v Otkrytykh Ruslah" (On the Computation of Unsteady Flow in Open Channels), *Meteorologiya i Gidrologiya*, 10:10-18, 1957.

Karlinger, M. R. and B. M. Troutman, "Assessment of the instantaneous unit hydrograph derived from the theory of topological random networks," *Water Resources Research*, 21(11); 1693–702, 1985.

Keulegan, G. H., "Spatially variable discharge over a sloping plane," *Trans. Amer. Geophys. Union.*, 25(6):956–968, 1944.

Karchenko, S. I. and S. Roos, "Experimental Investigations of infiltration capacity of watersheds and the prospects of accounting for the variability in the losses of rain water in computing flood runoff," *Soviet Hydrology*, Selected Papers, 6:537–554, 1963.

Kibler, D. F. and D. A. Woolhiser, "The kinematic cascade as a hydrological model," Hydrological paper 39, Colorado State University, 1970.

Kirkby, J. J., *Hillslope Hydrology*, Wiley, 1978.

——, "Hillslope Hydrology," *Hydrological Forecasting*, (ed. M. G. Anderson and T. P. Burt), pp. 37–75, Wiley, 1985.

Kirshen, D. M. and R. L. Bras, "The linear channel and its effect on the geomorphic IUH," *J. Hydrol.*, 65:175–208, 1983.

Kleiber, M., "Body Size and metabolism," *Hilgardia*. 6:315–353, 1932.

Klemes, V., "Physically based stochastic hydrologic analysis," *Advances in hydroscience*, II: 285–356, (ed. V. T. Chow), Academic Press, New York, 1978.

——, "Sensitivity of Water Resource Systems to Climate Variations," *World Climate Paper No. 98*, WMO, Geneva, 1985.

——, "Dilettantism in hydrology: transition or destiny?," *Water Resources Research*, 22(9):1775–1885, 1986.

Klute, A. and G. E. Wilkinson, "Some tests of the similar media concept of capillary flow. 1: reduced capillary conductivity and moisture characteristic data," *Soil Sci Sce. Amer. Proc.*, 22:278–281, 1958.

Kochel, R. C. and V. M. Baker, "Palaeoflood hydrology," *Science*, 215:353–36, 1982.

Kolmogorov, A. N., "Equations of turbulent motion of an incompressible fluid," *Izvestia AN SSR*, Ser. Fiz. 6:56–68, 1942.

Korzun, V. I., et al., *Mirovoi Vodnyi Balaons i Vodny Resursy Zemlii*, Gidrometeoizdat, Leningrad, 1974. Trans. into English, *World Water Balance and Water Resources of the Earth*,

Studies and Reports in Hydrology, No. 25, UNESCO, Paris, 1974.

Kozeny, J., "Flow in porous media," *S.B. Akad. Wiss. Wien. Abt.*, IIa, 127, 1928.

Kraijenhoff Van de Leur, D. A., "A study of nonsteady ground water flow with special reference to a reservoir coefficient," *Ingenieur*, 70(19): 87–94, 1958.

Launder, B. E. and D. B. Spaulding, *Mathematical Models of Turbulence*, Academic Press, 1972.

Leopold, L. B. and T. Maddock, *The hydraulic geometry of stream channels and some physiographic implications*, U.S. Geol. Survey Prof. Paper 252, 56 pp., 1953.

Luxmoore, R. J. and Sharma, M. L., "Runoff responses to soil heterogeneity: experimental and simulation comparisons for two contrasting watersheds," *Water Resources Research*, 6:675–684, 1980.

Machado, D. and T. O'Donnell, "A stochastic interpretation of a lumped overland flow model," *Modelling of hydrologic processes* (ed. Morel Seytoux), Water Resources Publications, Ft. Collins, Colorado, pp. 259–269, 1982.

Machmeier, R. E., "The effect of runoff supply rate and duration on runoff time parameters and peak outflow rates," Ph.D. thesis, University of Minnesota, University Microfilms, Ann Arbor, Michigan, 1966.

Machmeier, R. E. and C. L. Larson, "Runoff hydrographs for mathematical watershed model," *Proc. ASCE*, 94 HY 6:1453–1474, Nov. 1968.

____, *Muscles, Reflexes and Locomotion*, Princeton University Press, 1984.

Maller, R. A. and M. L. Sharma, "An analysis of areal infiltration considering spatial variability," *J. Hydrology*, 52:25–37, 1981.

Mandelbrot, B. B., *Fractals, Form, Chance and Dimension*, W.H. Freeman, San Francisco, 1977.

Mandelbrot, B. B., *The Fractal Geometry of Nature*, W. H. Freeman, San Francisco, 1983.

Mars Channel Working Group, "Channels and valleys on Mars," *Geol. Soc. Amer. Bull.*, 94:1035–1054, 1983.

McCarthy, G. T., *The Unit Hydrograph and Flood Routing*, U.S. Corps of Engineers, Providence, R.I., 1938.

McMahon, T. A., "Size and shape in biology," *Science*, 179:1201–1204, 1973.

Meier, M.F., "Snow and Ice in a Changing Hydrological World", *Hydrological Sciences Journal*, 28(1):3-22, 1983.

Meyer, O. H., "Simplified flood routing," *Civil Engineering*, 11(5): 306–307, 1941.

Miller, E. E. and R. D. Miller, "Physical theory for capillary flow phenomena," *J. Appl. Phys.*, 27:324–332, 1956.

Moore, R. J. and R. T. Clarke, "A distribution function approach to rainfall-runoff modelling," *Water Resources Research*, 17(5):1367–1382, 1981.

Moore, R. J., "The probability-distributed principle and runoff production at point and basin scales," *Hydrological Sciences Journal*, 30(2):272–296, 1985.

Morrison, P. and P. Morrison, *Powers of Ten*, Scientific American Books, New York, 1982.

Napiorkowski, J. J., "The optimization of a third order surface runoff model," *Scientific procedures applied to the planning, design and management of water resources systems,* Proc. of Hamburg Symposium Aug. 1983 (ed. E. Plate and N. Buras), IAHS Publ. No. 147, 161–172, 1985.

Nash, J. E., "Systematic determination of unit hydrograph parameters," *J. Geophys. Res.*, 64(1):111–115, 1959.

Nemec, J. and J. Schaake, "Sensitivity of water resource systems to climate variation," *Hydrol. Sci. J.*, 27(3):327–343, 1983.

Neuman, S. P., "Statistical characterization of aquifer heterogeneities: an overview," *Recent Trends in Hydrogeology* (ed. T. N. Narasimhan), Geol. Soc. Amer. Special Paper 189, pp. 81–102, 1982.

Nielsen, D. R. and J. Bouma, Soil Spatial Variability, Proc. of ISSS and SSSA Symposium, Las Vegas, Nov.–Dec. 1984, PUDOC, Wageningen, 1985.

Ol'dekop, E. M., *Ob isparenii s poverkh nosti rechnykh basseinov,* (On evaporation from the surface of river basins), Trans. Meteorol Observ. Iur-evskogo. Univ. Tartu, 4, 1911.

Peck, A. J., R. J. Luxmoore, and J. L. Stolzy, "Effects of spatial variability of soil hydraulic properties in water budget modelling," *Water Resources Research*, 13:348–354, 1977.

Peck, A. J., "Field variability of soil physical properties," *Advances in Irrigation,* Vol.2:189–22, 1983.

Philip, J. R., "An infiltration equation with physical significance," *Soil. Sci.*, 77:1153–1157, 1954.

———, "Field heterogeneity: some basic issues," *Water Resources Research,* 16(2):443–448, 1980.

Pike, J. G., "The estimation of annual runoff from meteorological data in a tropical climate," *Journal of Hydrology*, 2:116–123, 1964.

Polya, G., *How to solve it,* Doubleday and Co., Garden City, New York, 1957.

Popov, E. G., "Non-uniformity of surface retention as a factor of runoff," *Boll. Int. Assn. Hydrol. Sc.*, 7:21–26, 1962.

Prandtl, L., "Bericht uber Untersuchungen nur ausgebiedter Turbulenz," *S. Angew. Mat. Mech.*, 5:136–, 1925.

____, "Uber ein neuses Formelsystem fur die ausgebiedter Turbulenz," *Nachr. Ges. Wiss. Gottingen Math. Phys.K.*, 1:6–19, 1945.

Rodi, W., Turbulence Models and their Application in Hydraulics, Int. Assoc. Hydraulic Res., 1980.

Rodriquez-Iturbe, I. and Valdez, "The geomorphic structure of hydrologic response," *Water Resources Research,* 15(6): 1409–1420, 1979.

Russo, D., and E. Bressler, "Soil hydraulic properties as stochastic processes," *Soil. Sci. Soc. Amer. J.*, 45(4):682, 1981.

Rzhanitsyn, N.A., "Morphological and hydrological regularities of the structure of the river net," Trans. from Russian by D. B. Krimgold, U.S. Agric. Res. Service Dept, 56–92, 1960.

Saint-Venant, B. de, "Theorie du mouvement non-permanent des eaux," *C.R. Acad. Sci,* Paris, 148–54, 237–240, 1971.

Schaake, J. and Z. Kaczmarek, "Climate variability and the design and operation of water resource system," *Proc. of World Climate Conference*, 290–312, 1979.

Scheidegger, A. E., *The Physics of Flow through Porous Media*, University of Toronto Press, 1960.

____, "Statistical hydrodynamics in porous media," *Advances in Hydroscience* (ed. V. T. Chow), Academic Press, New York, 1964.

____, "Statistical theory of flow through porous media," Trans. *Soc. Rheol.*, 9:313–, 1965.

Schreiber, P., "Uber die Beziehungen zwischen dem Niedersclag und der wassfuhrung der Flusse In Mitteleuropa," Z Meteorol, 21(10), 1904.

Schumm, S. A., "The evolution of drainage systems and slopes in badlands at Perth Amboy, New Jersey," *Geol. Soc. Amer. Bull.*, 67:597–646, 1956.

Schumm, S. A. and R. W. Lichty, "Time, space and causality in geomorphology," *Amer. Jour. Science.* 263:110–119, 1965.

Schumm, S. A., "Palaeohydrology: application of modern hydrological data to problems of the ancient past," *IAHS Publ.* 185:93, 1967.

____, "Variability of the fluid system in space and time," *SCOPE Workshop on Spatial and Temporal Variability of Biospheric and Geospheric Processes*, St. Petersburg, Florida, Oct–Nov 1985.

Selvalingen, S., "ARMA and linear tank models," *Proc. of Third Fort Collins International Symposium*, (ed. H. J. Morel-Seytoux, J. D. Salas, T. G. Sanders and R. E. Smith), 297–313, 1978.

Sharma, M. L. and R. J. Luxmoore, "Soil spatial variability and its consequences on simulated water balance," *Water Resources Research*, 15:1567–1573, 1979.

Sharma, M. L., G. A. Gander, and C. G. Hunt, "Spatial variability of infiltration in a watershed," *J. of Hydrology*, 45:101–122, 1980.

Shreve, R. L., "Statistical laws of stream numbers," *J. Geol.*, 75: 179–86, 1966.

Sisson, J. B. and P. J. Wierenga, "Spatial variability of steady state infiltration rates as a stochastic process," *Soil. Sc. Soc. Amer. J.*, 45:669–704, 1981.

Smart, J. S., "Channel networks," *Advances in Hydroscience* (ed. V. T. Chow), Academic Press, New York, pp. 305–346, 1972.

Smith, R. E. and R. H. B. Hebbert, "A Monte Carlo analysis of the hydrologic effects of spatial variation of infiltration," *Water Resources Research*, 15:419–429, 1979.

Speakman, J. C., *Molecules*, McGraw Hill, New York, 1966.

Strahler, A. N., "Hypsometric analysis of erosional topography," *Geol. Soc. Am. Bull.*, 63:1117–1142, 1952.

Strupczewski, W. G. and Z. W. Kundzewicz, "Choice of a linear three-parametric conceptual flood routing model and evaluation of its parameters," *Acta Geophys. Pol.*, 28:129–1411, 1980.

Szesztay, K., "The hydrosphere and the human environment," Symposium on Representative and Experimental Basins, Wellington, New Zealand, December 1970.

Thompson, d' A. W., *On Growth and Form*, Cambridge University Press, 1917; second edition 1942; abridged edition 1961.

Thornes, J. B. and D. Brunsden, *Geomorphology and Time*, Halstead Press, Wiley, New York, 1977.

Torelli, L. and A. E. Scheidegger, "A review of the stochastic approach to flow through porous media," *Proc. of Conference on Stochastic Hydraulics* (ed. C-L Chiu), University of Pittsburgh, 1971.

Troendle, C. A., "Variable source area models," *Hydrological Forecasting* (ed. M. G. Anderson and T. P. Burt), Wiley-Interscience, 347–403, 1985.

Troutman, B. M. and M. R. Karlinger, "Unit hydrograph approximations assuming linear flow through topologically random channel networks," *Water Resources Research*, 21(5): 743–754, 1985.

Turc, L., "Le bilan d'eau des sols. Relation entre la precipitation, l'evaporation et l'ecoulement," *Ann Agron.*, 5:491–569 and 6:5–131, 1954, 1955.

Vauclin, M., G. Vachaud, and J. Imberon, "Spatial variability of some soil physical properties over one-hectare field plot," AGU Chapman Conference, Ft. Collins, Colorado, July 1981.

Vauclin, M., "Infiltration in unsaturated soils," *Fundamentals of Transport Phenomena in Porous Media* (ed. J. Bear and M. Y. Corapcioglu), Martinus Nijhoff, Dordrecht, pp. 257–313, 1984.

Viera, S. R., D. R. Nielsen, and J. W. Biggar, "Spatial variability of field measured infiltration rate," *Soil Sci. Soc. Amer.*, 45:1040–1048.

Warren, J. E. and H. S. Price, "Flow in heterogeneous porous media," *Soc. of Petroleum Engineers Journal*, 1:153–69, 1961.

Warrick, A. W., G. J. Mullen, and D. R. Nielsen, "Prediction of the soil-water flux based upon field-measured soil-water properties," *Soil Sci. Soc. Amer. J.*, 41:4–19, 1977.

Warrick, A. W., and A. Amoozegar-Fard, "Infiltration and drainage calculations using spatially scaled hydraulic properties," *Water Resources Research*, 15:1116–1120, 1979.

WCRP, Report of first session of JSC Scientific Steering Group on Land Surface Processes and Climate, *World Climate Paper* 96, WMO/ICSU, Geneva, January 1985.

Weinberg, G. M., *An Introduction to Systems Thinking*, Wiley-Inter-science, 1975.

Werrity, A., "The topology of stream networks," *Spatial Analysis on Geomorphology* (ed. R. I. Chorley), Harper and Row, New York, pp. 167–196.

Whitaker, S., "Diffusion and dispersion in porous media," *J. Amer. Inst. Chem. Eng* 13(3):420–427, 1967.

White, I., B. E. Clothier, and D. E. Smiles, "Pre-ponding constant-rate rainfall infiltration," *Modelling Components of Hydrologic Cycle* (ed. V. P. Singh), Water Resources Publications, Fort Collins, Colorado, pp. 127–148, 1982.

WMO, Intercomparison of conceptual models used in operational hydrological forecasting, *Operation Hydrology Report No. 7*, WMO, Geneva, 1975.

WMO, Outline Plan and Basis for the World Climate Programme

1980-1983, *WMO Publication No. 540*, WMO, Geneva, 1980.

WMO, Report on the International Conference on the Assessment of the Role of Carbon Dioxide and Other Greenhouse Gases in Climate Variations and Associated Impacts, *WMO Publication No. 661*, WMO, Geneva, 1986.

Wooding, R. A., "A hydraulic model for the catchment stream problem: 3 Comparison with runoff observations," *J. Hydrol.*, 4:21–37, 1966.

Youngs, E. G., "Use of similar media theory in infiltration," *Modelling Components of Hydrologic Cycle* (ed. V. P. Singh), Water Resources Publications, Fort Collins, Colorado, pp. 149–162, 1982.

Zavoianu, I., *Morphometry of Drainage Basins,* Elsevier, Amsterdam, 1985.

Zhao, R. J. and Y. Zhuang, "Regionalisation of the rainfall-runoff relations," *Proc. of East China College of Hydraulic Engineering,* (in Chinese), Nanjing, China, 1963.

Zhu, Yuan-Sheng, "The analysis of the relationship between the frequency of rainfall and annual flood series," U.S. China bilateral symposium on the analysis of extraordinary flood events, Nanjing, *Journal of Hydrology,* October, 1985.

JAMES C.I. DOOGE
University College, Dublin 2, Ireland.

Jim Dooge graduated from University College, Dublin, in 1942 with degrees of Bachelor of Engineering in Civil Engineering and Bachelor of Science in Mathematics, Mathematical Physics and Geology. Subsequently, he received the degree of Master of Engineering from the National University of Ireland in 1952 and the degree of Master of Science in Fluid Mechanics and Hydraulics from the University of Iowa in 1956.

From 1943 to 1946 he worked as a junior engineer on hydrometric survey and river improvement design with the Irish Office of Public Works. From 1946 to 1958 he worked on hydrologic and hydraulic design with the Irish Electricity Supply Board. In 1958, Jim Dooge was appointed as Professor of Civil Engineering at University College, Cork, and in 1970 was appointed to the same position in University College, Dublin. During the past thrity years he has been active in the promotion of new approaches to the study of hydrologic systems.

Jim Dooge was President of the Institution of Engineers of Ireland in 1968-69 and is currently President of the Royal Irish Academy. He was President of the International Association for Hydrological Sciences from 1975 to 1979 and was Secretary General of the International Council of Scientific Unions from 1980 to 1982. He has been, since 1980, Chairman of the Scientific Advisory Committee for the World Climate Impact Studies Programme and is Chairman of the organising committee for the Second World Climate Conference to be held in 1990.

Jim Dooge has received honorary doctorates from the Universities of Wageningen, Lund, Birmingham and Dublin. He was awarded the International Prize in Hydrology for 1983 of the International Association for Hydrological Sciences and the Bowie Medal for 1986 of the American Geophysical Union. He is a Foreign Member of the Polish Academy of Sciences and an Honorary Professor of the Technical University of Water Resources in Nanjing.

He was elected a member of the Irish Senate in 1961, was Deputy Chairman from 1965 to 1973, Chairman from 1973 to 1977, and Senate Majority Leader from 1983 to 1987. In 1981-1982 he was a member of the Irish Cabinet as Minister for Foreign Affairs.

FOREWORD

We have proposed elsewhere *) that *water resources,* as a scientific endeavor, be defined as the study of man's intervention in the hydrological cycle. Man intervenes in the hydrological cycle in order to satisfy individual and societal needs, whether to make water available for domestic, agricultural and industrial-commercial needs, or to use water bodies — static as well as flowing — for the removal of material and/or energy waste. In doing so, regional water resources may be depleted — particularly groundwater — and their physical, chemical and microbiological quality be impaired. The issue of managing water quality on a local, regional, national, continental and global scale is probably the most important problem facing today hydrologists and water resources specialists, as well as the public at large. The Sixth Kisiel Memorial Lecture addresses one of the most important aspects of this issue — that of groundwater contamination.

An analysis of this problem has two principal dimensions: a description of the (mainly) physical phenomena of fluid flow in aquifers and the attendant process of contaminant transport through the hydrogeological environment; a discussion of the socio-economic environment within which decisions related to managing groundwater quality are made. It is the latter that arouses considerable interest, since it involves a number of groups affecting the decision-making process. Some of these groups may be the waste generators (every individual generates waste), owner-operators of waste treatment and/or disposal facilities, insurance companies, Federal regulatory agencies, local governments, special interest groups, and others. Each of these groups views the problem of water quality management differently from the others, may or may not have a clear definition of objectives, and solutions proffered by one group could affect adversely other actors in this drama. Furthermore, the negotiations between these groups take place in an atmosphere where the unknowns seem to be more numerous than the known facts and the uncertainty is all-pervasive. Examples of vexing unknowns are (1) what is the effect of long-term exposure of humans to low doses of contaminants in concentrations at the limit of safety as established by regulatory agencies? (2) What are the synergistic effects of viruses and chemical contaminants on the public health of a region? Yet society cannot await results of long-term studies and investigations before making decisions regarding the maintenance of enviromental quality. Unfortunately, some environmental policies were formulated in the past without the benefit of the most recent knowledge

*) Foreword to the Fourth Kisiel Memorial Lecture.

and understanding of the physico-chemical and microbiological processes related to groundwater contamination. Clearly, such policies proved to be irrelevant.

The issue of managing water quality in general and more specifically the problem of groundwater contamination will be the subject of intense public debate. Scientific research, hopefully, will sharpen our perception of the alternatives available for maintaining a certain level of environmental quality, the societal costs involved in each alternative, the expected benefits, as well as the inherent risks. The Sixth Kisiel Memorial Lecture is a bright spotlight throwing considerable light on this complex issue.

Nathan Buras

GROUNDWATER CONTAMINATION: TECHNICAL ANALYSIS AND SOCIAL DECISION MAKING

R. ALLAN FREEZE

INTRODUCTION

I am honored to have been selected as the sixth Kisiel Lecturer. As time goes by, we must face the fact that the link between the Kisiel Lecturer and Chester Kisiel himself may become more distant. Perhaps, I will be one of the last who had a more-or-less direct link. Just before Chester's untimely death in 1973, he and I were planning a joint paper. It was not, however, a technical paper. We had discovered a mutual interest in the quotations of famous men and women that appeared to have special relevance to hydrology, and we were cogitating how we might pool our collections and share these delights with the hydrologic community. It seems that a reasonable time has finally come. Many of the quotations that accompany this essay are taken from Chester's collection.

I would like to think that he might also have been interested in the topic I have chosen. Hydrologists like Chester, whose primary interest lay in uncertainty and risk as it pertained to flood forecasts and other surface-water phenomena have long been familiar with the interplay between technical analysis and social decision-making. However, it is only in the last decade or so that this interplay has come into the domain of the groundwater hydrologist. And it has come with a vengeance! Groundwater contamination presents a threat to human health. To combat this threat, governments have enacted statutes and regulations. The existence of regulations means that engineering design must be carried out within a set of legal constraints. Because these legal constraints include economic penalties for noncompliance, there are economic constraints over and above those that apply to all engineering design. Moreover, because these legal and economic constraints are concerned with the protection of health and life, ethical questions arise. And lastly, because legal, economic and ethical questions are always subject to dispute, the entire process takes place in the political arena.

What is the role of technical analysis in all of this? How does the existence of a particular hydrogeological environment come into play in the engineering design of a particular waste-management facility or a particular remedial scheme, given the presence of a legal regulatory framework, economic constraints, and political realities? How does the technical analysis impact the social decision-making, or how should it? In this presentation, I will try to examine some of the interrelationships between design, prediction, uncertainty, risk, human error, regulation, ethics, equity, conflict, justice, and social decision-making.

Having immodestly laid down such a list, let me present the first of Chester's quotations:

> "I believe that a scientist looking at nonscientific problems is just as dumb as the next guy."
>
> Lee A. Dubridge

WASTE MANAGEMENT

Groundwater contamination can arise from both areal sources and point sources. In this presentation I will not discuss the problems associated with areal sources produced by agricultural application of herbicides and pesticides. Rather, I will emphasize point sources that may arise from waste management facilities. Among the types of facilities that have the potential to pollute groundwater are sanitary landfills for solid nonhazardous municipal waste, chemical landfills for solid and liquid hazardous industrial waste, tailings ponds for slurried mining wastes, sewage lagoons for liquid municipal waste, near-surface buried tanks for liquid industrial or low-level radioactive wastes, and deep repositories for solid high-level radioactive waste. Such facilities may be quite large, creating sources of a square mile or more; they are point sources only in the sense that they are concentrated and areally bounded.

In a recent report to the U.S. Congress, the Office of Technology Assessment (*OTA*, 1984) concluded that the actual and potential effects of groundwater contamination from waste-management facilities are significant and warrant national attention. They reported that cases of groundwater contamination have occurred in every state. They provided a list of substances that have been detected in contaminated groundwater. It includes over 175 organic and 50 inorganic chemicals, among which are liver and kidney toxicants, known or suspected carcinogens, and chemicals capable of damaging the reproductive and central nervous systems. Actual health damage requires that a pathway exist for the toxic chemical species from the source to the human organ. Determining this pathway involves understanding (1) transport through the hydrogeological environment, (2) concentration by biological agents in the biosphere, (3) human exposure through ingestion, inhalation or dermal contact, and (4) the toxicological principles of dose response. This presentation deals only with the first leg of such a pathway.

To combat the threat of increased groundwater contamination there has been a great deal of recent legislative and regulatory activity. There are currently 16 federal statutes in the United States that authorize programs relevant to groundwater protection (*OTA*, 1984), and all 50 states have programs in various stages of development. Most statutes and regulatory programs emphasize the reduction of contamination from waste-management facilities. The federal statutes of greatest significance are the Resource Conservation and Recovery Act (RCRA), under which new landfills are licensed, and the Comprehensive Environmental Response, Compensation, and Liability Act (CERCLA, popularly known as the "Superfund"), under which remedial action is mandated at sites that have failed.

In an economic sense, the need for regulation arises in response to "technical external diseconomies" (*Kneese and Bower*, 1968). Externalities result when contaminants discharged to the environment affect downstream users. Society is in general agreement that waste dischargers must not be allowed to neglect these offsite costs. Kneese and Bower note that market transactions are not likely to settle the externalities and that litigation is cumbersome. The environmental economics literature is rife with proposals for various types of economic incentives, marketable effluent-discharge permits and the like, but political acceptance has been slow to come, and few if any are in widespread use. In this presentation I will presume that direct governmental regulation will continue to be the most prevalent method of combatting environmental contamination.

There are at least three stages in the development and operation of a waste-management facility where there may be potential interactions between technical analysis and social decision-making: (1) during the siting process, (2) during facility design, and (3) during the design of remedial programs at facilities that have leaked. At each stage there are decisions that must be made between available alternatives. During the siting process the alternatives reflect the various overall waste management strategies that could be employed with respect to transportation, treatment, and method of disposal. Among the latter might be landfilling, containerization, or incineration. At the design stage, the various containment options must be weighed. For a landfill they might include natural or synthetic liners, leachate control systems, and capping. As pointed out by *Massmann and Freeze* (1987), containment is only part of the design process; it must be viewed in the context of the overall tradeoff that must be struck between site exploration, containment, and monitoring. At the remedial stage, where leakage has occurred, and a contaminated plume has already developed, alternative strategies include source control through excavation, capping, or slurry-wall construction, and plume control through pumping-well/injection-well scenarios.

PARTICIPANTS IN AN ADVERSARIAL ENVIRONMENT

> "A conservationist is someone who built his mountain cabin last year; a developer is someone who wants to build his mountain cabin this year."
>
> Old Colorado Saying

In this presentation, as in *Massmann and Freeze* (1987), I will explicitly recognize the adversarial relationship that exists in a regulated market economy between the various participants in the decision processes associated with the siting, design, and remediation of waste management facilities. The primary players are the owner-operator of the facility, and the government regulatory agency under whose terms the facility must be licensed. The owner-operator must decide which available alternatives are "best" on the basis of technical and economic criteria. The regulatory agency must decide which available alternatives are "acceptable" on the basis of regulatory statutes, social preferences, and political realities. The adversarial relationship need not be combative but it must be recognized that the objectives of each party are different and may in some sense be in conflict.

The owner-operator and the regulatory agency are not the only participants in the decision-making process. *Cantor and Knox* (1986) and *Kleindorfer and Kunreuther* (1985) have both identified a wider set of stakeholders. These include: (1) the waste generators, who may be held liable in the event of leakage, (2) insurance companies who hold insurance policies with waste-management firms, (3) local governments representing the host community, and (4) special interest groups representing agricultural interests, developer interests, property-owner interests, or environmental interests. While special-interest demands can be quite diverse, certainly the most common stands that arise during public participation involve opposition to a site or demands for remedial action.

Cantor and Knox (1986) have provided a classification of stakeholders. They divide them first into those affected by the problem and those affected by the solution. Each group is further divided into those that are directly affected and those that are indirectly affected. And lastly, the effect may be beneficial or it may be adverse. At the siting stage, the "problem" would be the inadequacy of waste-management capacity in a community, and the proposed solution would be a new facility. At the remedial stage, the problem would be a contaminant plume, and the proposed solution would be a particular remedial scenario. Stakeholders are "affected" due to proximity, economic impact, social impact, or values. In the latter two categories might be the perceived impact on community traditions and growth, and the personal values held with respect to economic development vis-a-vis environmental protection.

Like the owner-operator and the regulatory agency, each of these additional participants is attempting to determine what alternatives are "best" or "acceptable" given their own interests. I have placed the words "best" and "acceptable" in quotation marks; much of the rest of this essay will be concerned with how the various participants decide what is "best" or what is "acceptable" and how their deliberations impact on one another.

DECISION THEORY

> "Decision analysis is a formalization of common sense for decision problems that are too complex for informal common sense."
>
> Ralph L. Keeney

If we assume that each of the participants is acting rationally, then each of them is trying to maximize an objective function that represents their interests. The most general formulation of such an objective function is as a discounted stream of benefits, costs, and risks. In most risk-cost-benefit textbooks (*Crouch and Wilson*, 1982; *Lindley*, 1971) the objective function takes the form:

$$\Phi = \sum_{t=0}^{T} \frac{1}{(1+i)^t} \left[B(t) - C(t) - R(t) \right] \qquad (1)$$

where:
- Φ = objective function [$]
- i = discount rate [decimal fraction]
- t = time [yrs]
- T = time horizon [Yrs]
- $B(t)$ = benefits in year t [$]
- $C(t)$ = costs in year t [$]
- $R(t)$ = risks in year t [$]

The risks are further defined as:

$$R(t) = P_f(t)C_f(t)\gamma(C_f) \tag{2}$$

where: $P_f(t)$ = probability of failure in year t [decimal fraction]
$C_f(t)$ = cost associated with a failure in year t [$]
$\gamma(C_f)$ = utility function [decimal fraction, $\gamma \geq 1$]

In theory, all participants would calculate a value for Φ for each of the available alternatives, all of them using their own perceptions of the correct time horizon T and discount rate i, and all of them calculating the benefits, costs, and risks from their own perspective. In practice, some of the participants may carry out such an analysis tacitly; others may do so only by inference. In the following section, I will try to examine the groundwater contamination issue from the perspectives of the various participants using the terms that arise in equations (1) and (2). My use of this approach is not meant to suggest that I feel this is the only way to make rational decisions; it is simply a convenient framework from which to discuss the issues. In fact, there are many ethical dilemmas that arise in trying to place a dollar value on all the benefits, costs, and risks that arise; much of the later part of this presentation will be concerned with these dilemmas.

The risk term in equation (2) requires knowledge of the probability of failure of each alternative. The probability of failure obviously depends on the definition of failure, which will be different for each participant. The definitions of failure from the various perspectives will be included in the following section. Methods of calculation of the probability of failure will be briefly described for one of the participants in the later section on risk-based engineering design. The primary point to be made here is that calculation of the probability of failure constitutes the technical component of the decision analysis. It is the only component of the objective function that is based on engineering analysis.

The normalized utility function, $\gamma(C_f)$, that appears in equation (2) allows one to take into account the possible risk-averse tendencies of some decision makers. Figure 1 shows a utility function, $U_0(C_f)$, that represents the "expected-value" approach, and another, $U_1(C_f)$, that represents risk-averse behavior. The normalized utility function, $\gamma(C_f)$, is defined as:

$$\gamma(C_f) = \frac{U_i(C_f)}{U_o C_f)} \tag{3}$$

where $U_i = U_0$ for the expected-value approach, and $U_i = U_1$ for the risk-averse approach. For expected-value behavior, $\gamma = 1$ for all values of the cost of failure, C_f; for risk-averse behavior, $\gamma > 1$ for all C_f. The question of risk-averseness is addressed more fully in a later section.

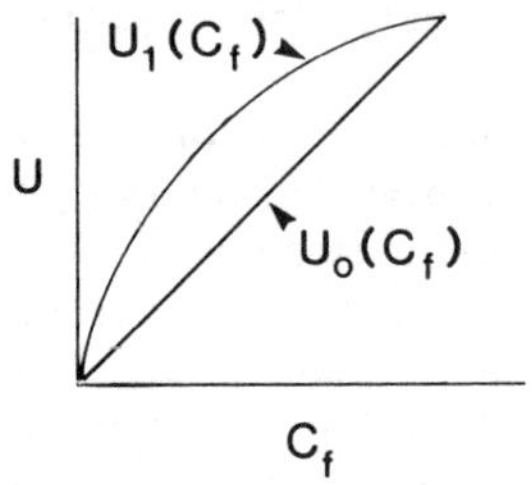

Expected-value:

$U_i = U_o$

$\gamma = 1$

Risk-averse:

$U_i = U_1$

$\gamma > 1$

$$\gamma(C_f) = \frac{U_i(C_f)}{U_o(C_f)}$$

In equation (1), the C(t) term represents actual costs, and the R(t) term represents probabilistic costs. One might ask why the actual benefits, B(t), are not balanced by a term for probabilistic benefits. Although I have not seen such a term in the risk-cost-benefit literature, I see real value to employing such a term for cases involving remedial alternatives. It would take the form:

$$V(t) = P_s(t)B_s(t)\gamma(B_s) \tag{4}$$

where: $P_s(t)$ = probability of success in year t [decimal fraction]
$B_s(t)$ = benefits associated with success in year t [\$]
$\gamma(B_s)$ = utility function [decimal fraction, $\gamma \geq 1$]

Use of the V(t) term requires a definition of success. In many cases, it may be that $F_s = 1 - P_f$, but it seems to me that this may not always be so. If there is a continuity of possible outcomes, there may be a certain level that would be defined as a success, which leads to otherwise unavailable benefits, and another level that would be defined as a failure, which leads to otherwise avoidable costs. If the V(t) term is incorporated, the objective function becomes:

$$\Phi = \sum_{t=0}^{T} \frac{1}{(1+i)^t} \{B(t) - C(t) + V(t) - R(t)\} \tag{5}$$

There is yet another extension to classical risk-cost-benefit analysis that I would like to invoke. It seems to me that it is not so much the absolute value of Φ associated with each alternative by each participant that drives social decision-making, but rather the change, $\Delta\Phi$, in comparison with the status quo. Before alternatives are floated and decisions broached, all of the participants are already experiencing a stream of benefits, costs and risks (as viewed from their own perspective). It is the potential change in these streams that drives social action. Under this extension, the objective function becomes:

$$\Delta\Phi = \sum_{t=0}^{T} \frac{1}{(1+i)^t} [\Delta B(t) - \Delta C(t) + \Delta V(t) - \Delta R(t)] \tag{6}$$

THE ADVERSARIAL DECISION PROCESS

> "Decide not rashly. The Decision made can never be recalled. The gods implore not, plead not, solicit not; they only offer choice and occasion, which once being passed return no more."
>
> Henry Wadsworth Longfellow

Let us look more closely at the terms in the objective function for three participants in the adversarial decision process: (1) the owner-operator of a waste-management facility, (2) a regulatory agency, and (3) a special interest group representing property owners or environmental interests. Table 1 summarizes these terms for decisions associated with siting and design; Table 2 does the same for decisions associated with remedial action. In Table 1 the participants are concerned with probabilistic costs associated with possible failure; in Table 2 they are hoping for probabilistic benefits associated with potential success.

Each group is likely to use a different discount rate, time horizon, and level of risk-aversion, as noted on Table 1. Direct benefits, B(t), represent income: revenues for services provided by the owner-operator, budget appropriations for the regulatory agency, and donations from supporters of the special interest group. The direct costs, C(t), are the outlays: for construction and operation of the waste-management facility by the owner-operator, for the administration of the regulatory agency, and for the mounting of a campaign by the special interest group.

It could be argued that the primary goal of special interest groups is to influence the regulatory agency. In this light, we will limit our further discussions to the two primary players on the stage — the owner-operator of a waste-management facility, and the regulatory agency that must license it.

For the siting and design of a new facility, both parties would probably be willing to define a "failure" in the same way. Following *Massmann and Freeze* (1987), I will define failure as a groundwater contamination incident that violates a performance standard established for the facility under existing regulatory policies. Presumably, a failure will be identified by the exceedance of a maximum permissible concentration for a particular chemical species in a regulatory monitoring well located at a compliance point.

Table 1 lists the probabilistic costs associated with failure for each of the participants. The owner-operator fears regulatory penalties, litigation costs, remedial costs, loss of goodwill, and the benefits foregone in the event of closure. The regulatory agency must consider the costs associated with the impairment of human health, and the failure to preserve clean water. These last two terms are not at all easy to evaluate and the next two sections identify the issues associated with attempts to place a dollar value on clean water and human life.

THE VALUE OF CLEAN WATER

> "We do not inherit groundwater resources from our parents; we borrow them from our children."
>
> Marcel Moreau
> Maine Department of Environmental Protection

The framework used in Table 1 presumes that societal interests with respect to groundwater contamination are in the hands of regulatory agencies. Regulatory officials must design policies under the direction of elected legislators that fulfil public desires. On Table 1, part of the cost associated with failure from the perspective of the regulatory agency is ascribed to the preservation of clean water.

It appears that there are two senses in which clean groundwater has value. It has value as a resource for the current generation, and it has value in storage for future generations. The first of these benefits can be evaluated with the concept of scarcity rent and the second in terms of preservation benefits.

Economists define rent as the difference between the benefits generated by a resource in its current use and the minimum sum the resource owner is willing to accept to keep the resource in its current use rather than divert it to an alternative use. The scarcity rent (*Howe,* 1979) of a depletable renewable resource is defined as the present value of all future sacrifices associated with the use of a marginal unit of the in-situ resource. Under appropriate market conditions, the scarcity rent is equal to the market value of these in-situ resources. At the prevailing market rates, the value of clean water in this sense is extremely low, and it is doubtful that economic justification for its preservation could be defended on these grounds.

By inference, the public must ascribe large values to one or more of the three components of preservation benefits: (1) option value, the protection of future options for competing usage, (2) existence value, a willingness to pay for the simple existence of clean water, and (3) bequest value, the satisfaction gained from bequeathing a clean natural environment to future generations. However, questionnaires completed by a random sampling of the public (*Greenley et al.*, 1982) do not confirm a widely-held consensus for large preservation values. *Raucher* (1984), having concluded that no form of remedial alternative could be economically justified at a particular contamination site using only direct benefits, back-calculated the value that preservation benefits would have to attain to justify remedial action. The numbers he obtained struck him as unreasonably large. They were orders of magnitude larger than those suggested by *Greenley et al.* (1982).

My interpretation of these and other similar findings is that clean water is not valued particularly highly by society. The public demand for groundwater cleanup as evidenced by the strong public support for the Superfund legislation is based on the desire for risk-reduction with respect to human health and life rather than benefit enhancement by the preservation of clean water.

THE VALUE OF LIFE

> "[The valuation of human life] is a procedure by which the higher is reduced to the level of the lower and the priceless is given a price. All it can do is lead to self-deception or the deception of others; for to undertake to measure the unmeasurable is absurd and constitutes but an elaborate method of moving from preconceived notions to foregone conclusions."
>
> E. F. Schumacher, in
> "Small is Beautiful"

In the usual approach to risk analysis, it is necessary to place a dollar value on human life so that it can be considered as one of the costs associated with failure from the regulatory perspective. *Fischoff et al.* (1981) provide a general summary of the methods that have been proposed to determine the value of a "statistical life," and *Sharefkin et al.* (1984) provide a discussion of the issue with particular reference to groundwater contamination. Most methods fall into one of the following classes: (1) the human productivity approach, based on the present worth of future lost earnings, (2) the legal approach, based on court awards for lives lost, (3) implicit valuation, based on the premium people are willing to pay to avoid increased risk, or that they demand when required to accept increased risk, (4) implicit valuation, based on observable responses to the risk associated with goods and services whose markets are reasonably well developed, and (5) implicit de-facto valuation, as embedded in government regulations already enacted.

This is an ethical quagmire, and I am inclined to agree with Schumacher that trying to measure the unmeasurable is absurd. Nevertheless it is clear that the primary burden placed on a regulatory agency by society is the preservation of human health and life, so comparison of the merits of alternative policies must be based on some measure that reflects their relative success in this area. *Massmann and Freeze* (1987) have chosen to use the total probability of failure, ΣP_f, as that measure, where:

$$\Sigma P_f = \sum_{t=0}^{T} P_f(t) \tag{7}$$

As they show, it is a surrogate for acceptable risk.

ACCEPTABLE RISK

The use of an acceptable risk criterion places decisions on the value of life into the political arena where they are resolved through the democratic process. As realized by *Baecher, et al.* (1980) and *Vanmarcke and Bohenblust* (1982) in their risk-based decision analyses of dam safety, this approach is equivalent to maintaining separate accounts for lives and dollars. For any given societal alternative, the economic costs and benefits are kept in one account and statistical lives saved or exposed in a separate account. When a particular project is put before the public for consideration, its acceptance or rejection is a measure of the politically acceptable limit of statistical lives, L_{pa}, that the public is willing to place at risk. If L is the best estimate of lives exposed by the alternative under assessment, then acceptance infers that:

$$\Sigma P_f . L \leq L_{pa} \tag{8}$$

If L is the same for each alternative in a set of alternatives, as it would be in assessing alternative policies for a given site, then (8) can be simplified to:

$$\Sigma P_f \leqslant (\Sigma P_f)_{pa} \tag{9}$$

where $(\Sigma P_f)_{pa}$ is the politically acceptable probability of failure. For any of the participants in the adversarial framework outlined on Table 1, equation (9) becomes a constraint that must be satisfied by the alternative that maximizes their individual objective functions. Of course, particular special interest groups may well attempt to influence the public perception of the correct value of $(\Sigma P_f)_{pa}$, and it is their democratic right to do so.

It is obvious that acceptable risk cannot be considered a fixed or known quantity. It is determined by the democratic process through elections, referendums, and public hearings, and under the influence of adversarial lobbies. It is dependent on values and beliefs. In practice, acceptable risk is the risk associated with the most acceptable decision; it is not acceptable in any absolute sense (*Fischoff et al.*, 1981).

RISK PERCEPTION AND RISK AVERSION

"Three-Mile Island is better built than Jane Fonda."

Bumper sticker in Richland, WA, site of the Hanford Nuclear Reservation

The risk-analysis literature abounds with articles about risk perception that show that people often do not have a realistic estimate of their risk due to actual or perceived threats. It would be easy to conclude from this data that people are stupid. I believe that a richer interpretation is provided by *Slovic et al.* (1980), who recognize that "riskiness" means more to people than simply "the expected number of fatalities." They suggest that attempts to characterize, compare, and regulate risks must be sensitive to the broader conception of risk that underlies people's concerns. The public apparently reacts to risks more strongly if they are involuntary, catastrophic, unfamiliar, uncontrollable, unobservable, or unknown to those exposed. They are particularly sensitive to ambiguity about risk.

Clearly, different stakeholders perceive risk differently, and it is to be expected that all of the participants in the social decision-making process will employ their own perceived risks in the calculation of their own objective function. Each individual par-

Table 1. Comparison of terms in the objective function for three participants in the adversarial decision process associated with siting and design of waste management facilities that have the potential to contaminate groundwater.

	Owner-Operator	Regulatory Agency	Property-owner and/ or Environmental Special Interest Group
Discount rate, i	Market interest rate (5 – 10%)	Social discount rate (2 – 5%)	Environmental discount rate (0%)
Time horizon, T	Engineering time horizon (10 – 50 years)	Social time horizon (100 – 200 yrs)	Environmental time horizon (200 – 10,000 yrs)
Direct benefits, $B(t)$	Revenues for services provided	Budget appropriation: siting & licensing	Donations from supporters
Direct costs, $C(t)$	Construction and operation of waste-management facility	Administration of regulatory agency	Cost of mounting campaign against site
Probability of failure, $P_f(t)$	Probability of groundwater contamination incident that violates performance standards at a compliance surface	Same as for owner-operator	Probability of facility being constructed at site
Probabilistic costs, $C_f(t)$ associated with failure	Regulatory penalties Litigation costs Remedial costs borne by owner-operator Benefits foregone Loss of goodwill	Impairment of human health or loss of human life Costs associated with failure to preserve clean water Litigation costs Remedial costs not borne by owner-operator	Reduction in property values Impairment of environment
Utility, $\gamma(C_f)$	Possibly risk-averse, $\gamma \geq 1$	Expected value, $\gamma = 1$	Risk-averse, $\gamma > 1$
Decision variables	Alternative designs for each potential site	Alternative regulatory policies, licensing requirements for each site	Alternative strategies

ticipant has the right to do so. There is a question, however, with respect to regulatory agencies. Should they use their "expert" perceptions of risk, or should they act as a conduit for public perceptions?

Risk aversion enters the analysis through the $\gamma(C_f)$ term in equations (2) and (3). Participants who have unrealistic risk perceptions may also be risk-averse. Among participants who have similar perceptions, risk aversion is largely controlled by the size of the penalty, C_f, associated with failure, and its relation to the net worth of the decision maker. Entities with a small net worth are likely to be more risk-averse than those with a large net worth. Risk aversion is also influenced by the availability of liability insurance. Regulatory agencies presumably ought to make decisions in an expected-value climate without risk aversion.

So what about the current climate surrounding groundwater contamination and waste management? My own view is that the great majority of the public has a relatively accurate perception of the risks. I believe that the underestimation of the potential health risks by professionals in the waste-management industry is at least as serious a matter as the overestimation of the risks by extreme environmental groups. There is much ambiguity about the toxicological data and the long-term effects of low doses of organic contaminants and radionuclides. Under these circumstances, a degree of risk aversion is not irrational.

In any case, in our democratic society, if large numbers of the public believe something to be true, it is right and proper for public policy to take this belief into account, even if that belief is at odds with prevailing technical analysis. It is the responsibility of the technical community to educate the public where its technical views are naive. However, it must be remembered that the technical community also has diversity with respect to values and political beliefs. There is a fine line between education, advocacy, and propaganda.

UNCERTAINTY

> "If a man will begin with certainties, he shall end in doubts; but if he will be content to begin with doubts, he shall end in certainties."
>
> Francis Bacon

The risk terms in the objective function include the probability of failure, $P_f(t)$, and/or the probability of success, $P_s(t)$. This introduction of a probability measure is tacit recognition of the role of uncertainty in the decision-making process. Moreover, because the $P_f(t)$ term is the technical component of the risk-cost-benefit analysis, it is the uncertainty in prediction that is being recognized.

Earlier, I defined failure as a groundwater contamination incident that violates a performance standard at a compliance point. In this context, a failure involves two independent sequential events: (1) breaching of the containment structure, and (2) migration of the released contaminants through the subsurface

Table 2. Comparison of terms of the objective function for three participants in the adversarial decision process associated with remedial action at a waste-management facility that has contaminated groundwater.

	Owner-Operator	Regulatory Agency	Property-owner and/ or Environmental Special Interest Group
Direct benefits B(t)	None	Budget appropriation: remedial action	Donations from supporters
Direct costs C(t)	Remedial costs borne by owner-operator	Remedial costs not borne by owner-operator Cost of administration of remedial action	Cost of mounting campaign demanding cleanup
Probability of success, $P_s(t)$	Probability of remediating site to meet performance standards	Same as for owner-operator standards	Same as for owner-operator, but with possible added desire to close site
Probabilistic benefits, $B_s(t)$, associated with success	Reinstigation of direct benefits Avoidance of further penalties and litigation	Removal of threat to human health Reclamation of clean water	Removal of threat to human health Recouping of property values
Decision variables	Alternative designs for remedial action	Alternative cleanup standards	Alternative strategies

hydrogeologic environment to the compliance point. There will be uncertainty with respect to the prediction of both these events. If we assume that the predictions are based on a computerized numerical mathematical model based on a boundary-value problem that describes the physical system, then our uncertainty may arise from any or all of three sources: (1) model error, (2) errors in the boundary or initial conditions, and (3) parameter error. For the migration component of the potential failure mechanism the third source is thought to be much more important than the first two. The parameters of interest are those properties of the porous medium that control the transport of contaminants: hydraulic conductivity, porosity, dispersivity, diffusion coefficient, and retardation factor. For any realistic geologic environment these parameters can be expected to exhibit heterogeneity through space. There is, of course, only one spatial distribution for each parameter that actually exists at a given site, but because it is impossible to measure the parameters at all points in the system, there is uncertainty as to the actual value at any unmeasured point. In recent years, hydrogeologists have learned how to represent these uncertain heterogeneous parameter distributions using the theory of spatial stochastic processes and the tools of geostatistics (*Neuman*, 1982; *de Marsily*, 1984). Solutions to the stochastic boundary-value problems lead to probability density functions for travel times or mass fluxes. Such output is well suited to risk analysis and risk-based engineering design.

The application of stochastic simulation at field sites is best carried out in a framework in which the estimates of the moments of the probability density functions of hydrogeologic parameters are updated, and predictive uncertainties of output travel times are reduced, through the collection of additional field measurements at the site. This Bayesian approach was pioneered by *Hachich and Vanmarcke* (1983) and has recently been applied by *Massmann and Freeze* (1987).

Consider the probability density function for contaminant travel times shown in Figure 2. Assume that the probability of failure, P_f, is defined as the probability that the travel time, t, is less than t*. We can reduce P_f in one of three ways: (1) by making additional measurements of hydrogeological parameter values in order to reduce our uncertainty, σ_t, on travel time, (2) by adding engineered barriers to the design of the waste-management facility in order to increase the mean travel time, $\bar{t}$, or (3) by installing a monitoring system at a position between the source and the compliance point, thus providing an opportunity to remediate a leak before it becomes a failure. It is the role of the owner-operator's decision process to choose the "best" mix of these three options from his perspective. It is the role of the regulatory agency to place licencing restrictions on the owner-operator such that his mix will be politically-acceptable.

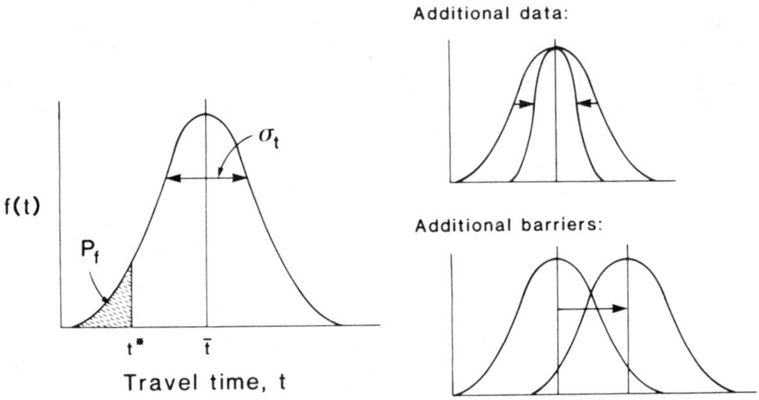

HUMAN ERROR

"It ain't so much the things we don't know that get us in trouble. It's the things we know that ain't so."

Artemus Ward

The probability of failure that is calculated with stochastic simulation reflects the uncertainties in the technical data. It is a theoretical probability of failure. However, post-facto studies of engineering failures often point to human error as the source of failure rather than technical uncertainty. The well-documented groundwater contamination incident at the Hanford Nuclear Reservation due to leakage from Tank 241-T-106 in 1973 (*Routson et al.*, 1979), for example, has been ascribed to a breakdown in the administration of the monitoring network when an employee went on vacation.

To the best of my knowledge, the probability of failure due to human error has not come under consideration in the design of waste-management facilities or their remediation. However, the question has been considered by structural engineers. *Ditlevsen and Hasofer* (1984) suggest that alternative designs be classified with respect to "mistake-robustness." They recommend optimizing first with respect to the theoretical probability of failure, $P_f(t)$,

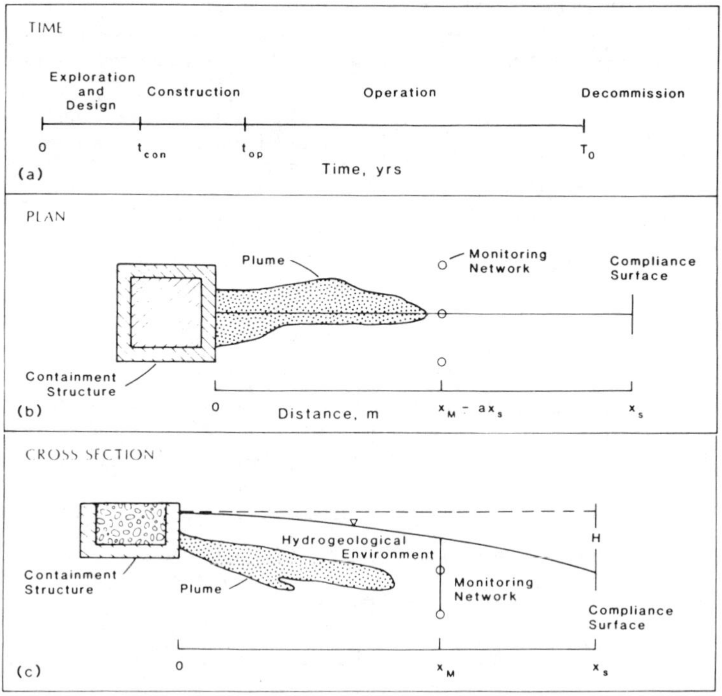

within each mistake-robustness equivalence class, and then optimizing with respect to $P_e(t)$, the probability of failure due to human error. Perhaps there are applications of these techniques in the design of facilities that may pollute groundwater. The analysis presented in the next section does not include consideration of human error.

RISK-BASED ENGINEERING DESIGN

> "Nature is indifferent to the difficulties it causes a mathematician."
>
> Fourier

In this section I will briefly summarize some of the results presented by *Massmann and Freeze* (1987). They presented a method of risk-based engineering design carried out from the owner-operator's perspective. Table 3 summarizes the types of decision variables faced by the owner-operator at each stage of development of a waste-management facility. Massmann and

Table 3: Decision variables for owner-operator.

Siting Stage	Site Selection	Number and Location of Sites to Investigate
Design Stage	Site Investigation	Number, location and depth of drillholes. Number, location and depth of parameter measurements. Parameters to measure.
	Containment	Number, thickness and permeability of synthetic liners. Number and location of drains and/or wells for leachate collection.
	Monitoring Network	Number, location and depth of monitoring points. Species to be analyzed. Frequency of sampling.
Remedial Action Stage	Plume Investigation	Number, location and depth of drillholes. Number, location and depth of sampling points. Species to be analyzed.
	Source Control	Location, depth, and permeability of slurry well. Thickness and permeability of cap.
	Plume Control	Number, location and depth of pumping and injection wells. Pumping and injection rates.

Freeze limit their considerations to the design stage. The design alternatives involve various apportionments of the available budget between site investigation, containment, and installation of a monitoring network. In their study, the facility is a landfill and containment is to be achieved through one or more synthetic liners. The geometry is shown in Figure 3. They assessed a hypothetical hydrogeological environment consisting of a surficial aquifer of unconsolidated sand and gravel. The site investigation is limited to the collection of subsurface hydraulic conductivity measurements from the aquifer. The flow system is assumed to be steady-state and fully saturated. Transport is limited to a single, inorganic, non-radioactive species. Under these circumstances advection is the primary transport mechanism, and dispersion and retardation will have only a secondary influence.

To analyze this system, Massmann and Freeze put together a design scheme that integrates several techniques into a risk-cost-benefit framework. They calculate the probability of failure using: (1) reliability theory with an exponential probability density function for the breach times of the synthetic liners, (2) a two-dimensional finite-element aquifer model, (3) a geostatistical description of the hydrogeological environment with a lognormally-distributed, exponentially-autocorrelated representation of hydraulic conductivity, and (4) a Monte Carlo conditional stochastic simulation. The worth of additional data is assessed with a Bayesian updating scheme after *Hachich and Vanmarcke* (1983).

The approach they followed was to set up a base case and then perform sensitivity analyses to examine the effect of alternative design options on the owner-operator's objective function. The base case is a midsized landfill with an area of 30,000 m^2, a capacity of 450,000 tons, and a throughput of 45 tons/day. The mean hydraulic conductivity $\bar{K}$ of the aquifer was taken as $5x10^{-3}$ cm/s. The aquifer has a depth of 10 m, and there is a gradient of 0.008 over the 1000 m distance between the source and the compliance point. The base case uses a single synthetic liner with a mean breach time of 15 years. Site investigation involved 3 hydraulic conductivity measurements, and there are also 3 monitoring points in the base-case. The owner-operator uses a discount rate of 10% over a 46-year time horizon. The charge for waste handled is set at $90/ton. The probabilistic costs in the event of failure total $15 million. The base case produces a rather complex stream of benefits, costs and risks; the value of the owner-operator's objective function, Φ, is $1.1 million.

Table 4 shows the results of sensitivity analyses on three decision variables. In Table 4a the effect of the number of liners is shown. Additional liners decrease the probability of failure and reduce risk. Table 4a shows that this risk reduction has greater benefit than the cost of the additional liners, and the value of Φ increases as additional liners are installed. If the owner-operator is a rational decision maker, he will use two liners rather than one or zero.

Table 4b shows that for the particular case presented, the risk reduction afforded by a more closely-spaced monitoring network just barely recovers the costs incurred for installation. The almost equal values of Φ for the three cases would lead a rational

Table 4: Results of sensitivity analysis on three decision variables from the owner-operator's perspective.

(a) Containment		Base Case	
No. of liners	0	1	2
Φ	$\$-0.26 \times 10^6$	$\$1.1 \times 10^6$	$\$1.9 \times 10^6$
ΣP_f	0.81	0.64	0.06

(b) Monitoring		Base Case	
No. of monitoring points	0	3	11
Φ	$\$1.2 \times 10^6$	$\$1.1 \times 10^6$	$\$1.1 \times 10^6$
ΣP_f	0.79	0.64	0.33

(c) Siting		Base Case	
$\bar{K}$	5×10^{-2}	5×10^{-3}	5×10^{-4}
Φ	$\$0.1 \times 10^6$	$\$1.1 \times 10^6$	$\$2.1 \times 10^6$
ΣP_f	0.74	0.64	0.5×10^{-5}

designer to be indifferent to the monitoring option, and he might well choose not to monitor.

Table 4c shows the importance of siting to the owner-operator. For a fixed containment design, as the hydraulic conductivity decreases (i.e., a more favorable site is selected), the value of the objective function increases.

Let me emphasize that Table 4 is simply an example of the methodology. It would be easy to construct hypothetical cases where containment is less valuable or monitoring more valuable. It is not intended that the reader draw generalized conclusions from this single example.

Table 5: Decision variables for regulatory agency.

Siting Stage	Site Selection	Site Selection Acceptability Criteria
Design Stage	Compliance Surface	Location. Frequency of sampling.
	Performance Standards	Maximum permissible concentrations.
	Design Standards	On liners: Number, thickness and permeability. On leachate collection: Number & location of drains. On monitoring: Number and spacing.
	Penalties	Amount of performance bond. Amount and timing of fines. Conditions for closure.
Remedial Action Stage	Design Standards	On source control measures: Depth and permeability of slurry wall. On plume control measures: Number and spacing of wells.
	Performance Standards	Maximum permissible concentrations after remediation.
	Cost Recovery	Source investigations. Litigation strategy.

REGULATORY ISSUES

"Social legislation cannot repeal physical laws."

D. P. Oaks, President
Brigham Young University

The primary issue in this essay is societal risk, not owner-operator risk, so let us now switch hats and look at things from the perspective of the regulatory agency. As noted earlier, the direct application of a regulatory risk-cost-benefit analysis founders on the difficulty of assigning dollar values to clean water and human health. However, the regulatory agency can learn much by examining the impact of alternative regulatory policies and licencing requirements on the owner-operator's design strategy. Table 5 summarizes the decision variables for the regulatory agency. *Massmann and Freeze* (1987) assess the question of the relative worth of performance standards and design standards at the design stage.

Recall that we earlier recognized that the total probability of failure, ΣP_f, taken over the time horizon of the facility, is a surrogate for acceptable risk. On Table 4a it can be seen that ΣP_f decreases as additional liners are installed. The value $\Sigma P_f = 0.06$ for the two-liner case might be politically-acceptable. In the case shown, the interests of the owner-operator and the interests of the regulatory agency are compatible; both would favor the two-liner design. However, this is not always the case. As shown on Table 4b, for example, the regulator would favor the denser monitoring network, the owner-operator, the sparser one. In such cases, design standards would have to be imposed by the regulatory agency if it is desired to force an owner-operator to use a design that reduced risk to society.

Performance standards are usually coupled with fines or other penalties for noncompliance. Such prospective fines to be imposed at some future date in the event of failure have little impact on the design decisions made by the owner-operator at the time of facility construction due to the influence of the discount rate. The fines may recoup the costs of enforcement and remediation but they do not lead to designs that reduce risk to society. For these reasons, it is my opinion that design standards have greater potential to protect society than performance standards. There is a down side, however: design standards tend to discourage innovative design. To avoid this, regulatory agencies have to be willing to apply design standards with a flexible, site-specific approach.

Table 4c shows that if we really want to reduce ΣP_f, the answer lies in siting. The only value of ΣP_f on Table 4 that can be clearly identified as politically acceptable is associated with the lowest permeability site. Regulatory control of siting would be more effective than any other type of regulatory practice. Unfortunately, it is hard to envisage the removal of the siting process from the political arena.

ECONOMIC ISSUES

> "You have only to take in what you please and leave out what you please, multiply and divide at discretion, and you can pay the National Debt in half an hour. Calculation is nothing but cookery."
>
> Lord Brougham

It is widely recognized that decisions based on an objective function grounded in risk-cost-benefit analysis are very sensitive to the value of the discount rate. For the owner-operator there is little mystery. If he borrows money at the market interest rate to invest in the construction of the waste management facility, then he will use this interest rate as the discount rate in his objective function. If this rate is 10%, the net present value of any economic event that occurs more than 25 years into the future will approach zero and it will have no impact on his objective function. No matter what technical time horizon may be specified, his economic time horizon will be 25 years or less. As a result, regulatory penalties to be imposed at the time of failure sometime in the distant future will have little impact on his current decision-making with respect to siting or design. Clearly, there is a strong dichotomy between these short time horizons and the much longer time horizons over which the impacts of social decisions ought to be considered.

The need for longer time horizons for social decisions implies the need for lower discount rates. If the future generations are to be protected from the risks associated with current decisions, an expected-value approach with a discount rate of zero would be required. There are many arguments in the economics literature for a social discount rate that is lower than the market rate (*McDonald*, 1981) but only the environmentalists argue for a zero rate. Even low non-zero discount rates do not lead to economic time horizons on the order of those suggested for hazardous and nuclear waste.

Paté-Cornell (1984) makes an interesting argument in this regard. She argues that the problem lies not in the specification of the discount rate and the time horizon but in the suitability of risk-cost-benefit analysis. She states that if irreversible damage will occur to future generations due to a policy then it should also be unacceptable to the current generation, and risk-cost-benefit analysis is not a suitable tool. If one decides that for ethical or political reasons, it is not appropriate to balance risks, costs, and benefits today, this also applies to tomorrow. Conversely, current generations have the right to use resources and create risks for the future, but only to the extent that they would do the same if the risks were to be incurred today. If economic efficiency has been judged relevant in this light, then risk-cost-benefit analysis is appropriate and it should be carried out with an appropriate discount rate.

If I have interpreted these arguments correctly, I believe that they provide further support for the need for an acceptable-risk constraint determined in the political arena. We must trust that a component in the political judgment leading to the acceptance of a risk will be consideration of future generations.

EQUITY AND JUSTICE

> "Economic decisions have a large zero-sum element. On average, society may be better off, but this average hides a large number of people who are much better off and a large number of people who are much worse off. If you are among those who are worse off, the fact that someone else's income has risen by more than your income has fallen is of little comfort."
>
> Lester C. Thurow in
> "The Zero-Sum Society"

A decision that is optimal in some sense for society as a whole will not be optimal for all the individuals in the society. As *Thurow* (1980) recognizes in his concept of the zero-sum society, there are always winners and losers. This lack of equity may occur temporally or spatially. Temporal inequity, which was discussed in the previous section, involves the postponing of risk to future generations. Spatial inequities lead to LULU's ("locally unwanted land uses"); and the NIMBY syndrome ("not in my back yard"). As *Willard and Swenson* (1984) have pointed out, people resist LULU siting because they fear disease and death. These fears are not irrational, and policy makers must accept people's fears as a valid part of policy-making.

The NIMBY syndrome is not a simple one to solve. If one believes in the risk-cost-benefit approach, then the obvious solution requires that benefits be offered to a community that hosts a waste management facility in an amount equal to their costs and risks. This simple solution hides many complexities. Host communities consist of many individuals who differ in their politics, values, and economic status. Some individuals would be harmed by a waste-management facility; others would stand to gain. What form should the benefits take? How should the amount of compensation be determined? One suggestion (*Kleindorfer and Kunreuther*, 1985) is that host communities bid for waste-management facilities, with the community that demands the least compensation being selected. This approach would internalize the political struggle from that of repelling an external intruder to conflict within the host community. It is not clear that this is necessarily a desirable social outcome.

The question of what constitutes justice in these cases is really a philosophical issue (*Paté-Cornell*, 1984). One either believes in libertarian principles, which emphasize individual rights; egalitarian principles, which favor the least well-off in society; or utilitarian principles, in which decisions are made to maximize the sum of individual utilities. Risk-cost-benefit analysis, and most practical decision-making, falls into the utilitarian framework. It leads to unavoidable inequity and a zero-sum society with winners and losers.

In a democratic society, the losers will be those who are politically weak. In a capitalist society, those who are politically weak are those who are economically weak. In short, those who are least well-off will be asked to bear the additional probabilistic costs associated with risk. The pessimistic conclusion of this section is that if we view the purpose of social-welfare schemes to be some measure of income redistribution, then the current methods of siting and regulating waste-management facilities are not a part of the social welfare program.

CONFLICT

> "Purposes, as understood by the purposer, will be judged otherwise by others."
>
> F. P. Chisolm, in "Basic Laws of Frustration, Mishap and Delay"

> "Where there's smoke, there's usually a smoke-making machine."
>
> John F. Kennedy

Conflict takes place between individuals. *Cantor and Knox* (1986) identify four sources of conflict that may arise in siting or remediating waste-management facilities. There may be cognitive conflict (over facts), values conflict (over goals), interest conflict (due to inequity), or relationship conflict (due to personalities or political stance). The potential for conflict resolution through public participation, negotiation, and other types of formal and informal interaction, is heavily dependent on the type of conflict. Conflicts over technical facts and interpretations are usually resolved to a large degree prior to decision-making. Conflicts over inequities or values are largely unresolvable. In cases involving serious groundwater contamination, all four types of conflict usually arise in a complex public stew stirred by the media. Ultimately, politically acceptable decisions are taken, with the usual fallout of technical, economic, and political winners and losers.

If we view the fundamental conflict as one between the waste-management firms, who contribute to the economic health of society, and government regulatory agencies, who protect societal interests with respect to health and environment, then it is incumbent on us to ensure that the two sides are evenly matched. I am not qualified to judge the economic and political skills of the respective sides, but in terms of technical expertise, which ought to be a primary foundation for this type of social decision-making, it is my fear that the two sides are not at parity. Because of poor pay scales and bureaucratic frustration, the technical skills on the regulatory side of the negotiation table are much weaker than those on the side of the waste-management firms (and their engineering consultants). Young regulators with good technical skills are quickly lured away from the agencies into the consulting world. In the long run, I fear that this inequality may lead to decisions that could result in a higher level of groundwater pollution than is mandated in the statutes that the regulatory agencies are set up to administer.

ETHICAL ISSUES

> "The means by which we live have outdistanced the ends for which we live. Our scientific power has outrun our spiritual power. We have guided missiles and misguided men."
>
> Martin Luther King

There are a host of ethical dilemmas that can arise for the various participants during the social decision-making process associated with potential or actual contamination from waste-management facilities. In this section, I will introduce four such issues.

The first concerns the role of the design engineer in protecting public health and safety. In Table 1 I have placed this entire responsibility in the hands of the regulatory agency, and have assumed that design engineers will not concern themselves with this issue if an adequate regulatory system is in place. The cost of failure in the risk term for the owner-operator, who the design engineer serves, involves potential regulatory penalties, not loss of life. Engineers function under a code of ethics: the first Fundamental Canon in the Code of Ethics of the American Society of Civil Engineers (*Firmage*, 1980) states that engineers shall hold paramount the safety, health and welfare of the public in the performance of their professional duties. In the absence of regulations, the design engineer would presumably prepare designs for a waste-management facility that are in keeping with his interpretation of the code of ethics. However, if the regulatory system in place has the respect of the engineering community, I believe that design engineers will feel they have satisfied their ethical obligations if they meet the regulatory requirements. There is no question that this is an accurate reflection of existing practice in the nuclear-waste and hazardous-waste industries. Furthermore, I believe that such behavior is ethical. The point is that an engineering ethic cannot be viewed as morally absolute; it is a function of the regulatory climate.

I would now like to describe three types of behavior which I believe to be unethical and therefore against the public interest.

Looking first from the perspective of the owner-operator, it would be possible for a firm to include in their risk term not only the probability of failure, but also the probability of detection of a failure by the regulatory agency, the probability of enforcement if detected, and the probability of conviction if the case is taken to a court of law. Interestingly enough, economists are not blind to this possibility. *Harford* (1978) discusses the behavior of firms under imperfectly enforceable pollution standards. If not morally constrained, it can be to the firm's economic advantage to violate pollution control laws. Firms may also adopt various kinds of strategic behavior to affect the penalty structure of the regulatory agency. They may threaten to go out of business or to move to an area where the agency does not have authority. Firms might collude to violate pollution control laws simultaneously, thereby overloading the agency's ability to enforce its laws effectively.

Regulatory agencies are not free from unethical behavior. They deal with a large number of contamination incidents and they encounter a wide spectrum of cooperation, or lack thereof, from the owner-operators in the negotiations leading to remedial action. In general, regulatory agencies have not been willing to identify cooperative firms to the public for fear of being accused of being in bed with the polluters. The result is that potentially harmonious and efficient cleanup procedures are turned into acrimonious conflicts for political reasons. If the firms perceive that there is no advantage to cooperation, even the good corporate citizens may not do so. Dishonesty on the part of the regulatory agency as to the true state of affairs is unethical and it has a social cost.

Lastly, I note that there can be a level and style of advocacy on the part of special interest groups that can be unethical. If the true aims of the group are different from the stated ones; if facts that are known to be false are used in arguments; in short, if the ends justify the means, then such advocacy is unethical.

One would like to think that unethical behavior receives its just desserts. In the long run, I suspect that this is so, but in the short run, unethical behavior has great potential to play havoc with our system of social decision-making. I have no solutions to suggest, other than to add my voice to the growing clamor for a return to a higher moral standard in everyday business affairs. The decline in ethical standards in that community is affecting society on a much broader front than is commonly recognized.

CRIME

Rothermal (1983) has argued that it is in the best interests of society that the hazardous-waste management business be profitable. If it is not, many companies that are capable of providing technical skills needed for hazardous-waste disposal will avoid this opportunity because of its high visibility and the associated notoriety in the event of accidents. If reputable companies avoid the field, the demand will be filled by disreputable companies. Brown (1979) documents a variety of illegal activities associated with waste management. They range from small-scale "midnight haulers" to large-scale infiltration of the industry by organized crime. If enforcement of environmental statutes is uncertain due to underfunding, then illegal suppliers of waste-management services can thrive.

The challenge to the regulatory agencies is great. They must put in place a set of regulations and an enforcement procedure that protects the health and safety of the public and discourages the criminal element, yet allows for a healthy return on investment by reputable businesses. If the public, through the political process, does not support a sufficient level of funding for the agencies, the agencies will be forced to choose between a system of slack enforcement of stringent regulations, or strong enforcement of weak regulations. The first will lead to environmental pollution from illegal sources; the second will lead to pollution from poorly designed facilities.

SUMMARY

It is not my intention to provide a detailed summary of the material I have presented in this essay. Rather, I would like to highlight a few of the ideas that I think have particular importance at the boundary of technical analysis and social decision-making.

1. The role of technical analysis lies in providing estimates of the probability of failure (of new facilities) or the probability of success (of remedial action) as a component in the risk term of the objective functions for the various participants in the social decision-making process.
2. Strong public support for groundwater remediation at sites that have been contaminated apparently rests on a desire for risk reduction with respect to health concerns rather than concern for the preservation of water resources.
3. Ultimately, these concerns are settled in the political arena through the determination of a politically acceptable risk.
4. There is a temporal inequity caused by the disparity between the discount rates and time horizons used by owner-operators of waste-management facilities and those used by a regulatory agency representing societal concerns with respect to the potential health risks. It is to be hoped that the political judgment leading to an acceptable risk includes consideration of the risks to future generations.
5. Public risk aversion to the potentially adverse health effects of contaminated groundwater is not irrational. It is right and proper for government agencies to take public risk perceptions and risk aversion into account in setting regulatory policy.
6. Regulatory agencies are more likely to induce engineering designs that lead to low societal risk if they impose design standards rather than performance standards.
7. Regulatory control on siting would be far more effective than any other form of regulatory practice in reducing societal risk.
8. We live in a zero-sum society. Without specific regulatory protection, the siting of waste-management facilities leads to unavoidable inequity that usually falls on those that are weakest politically and economically.
9. Risks to society increase if the adversarial conflict between owner-operators and regulatory agencies is not evenly matched.
10. It is unethical for regulatory agencies to let political concerns keep them from identifying good corporate citizens during site remediation.
11. Insufficient political support for proper regulatory funding will lead to higher societal risks.
12. Regulatory practice must not preclude the existence of an economically healthy waste-management industry.

Let me close with a final gem from Chester Kisiel's list of quotations. It seems to have special relevance to the application of science and engineering in social decision-making:

> "There is something fascinating about science. One gets such wholesale returns of conjecture out of such a trifling investment of fact."
>
> Mark Twain

ACKNOWLEDGEMENTS

I hope my many references throughout the text to *Massmann and Freeze* (1987) will make clear my indebtedness to Joel Massmann. Many of the ideas in this essay grew out of the interactions I had with Joel over the past three years.

I also want to thank my colleagues at the University of Arizona who provided a stimulating environment to Joel and me during my sabbatical year in 1984–85 when the work on which this article is based began: Tom Maddock, Shlomo Neuman, Soroosh Sorooshian, Don Davis, Nathan Buras, Lucien Duckstein, and Sid Yakowitz.

I have also benefited from discussions on these topics with Leslie Smith, Steve Burges, Ron Cummings and Dennis McLaughlin.

REFERENCES CITED

Baecher, G. B., M. E. Pate and R. de Neufville, Risk of dam failure in benefit-cost analysis, *Water Resour. Res., 16*, 449–456, 1980.

Brown, M., *Laying Waste: The Poisoning of America by Toxic Chemicals*, Pantheon Books, 1979.

Cantor, L. W. and R. C. Knox, *Ground Water Pollution Control*, Lewis Publ. Inc., 526 pp, 1986.

Crouch, E. A. C., and R. Wilson, *Risk/Benefit Analysis*, Ballinger, 1982.

Ditlevsen, O. and A. M. Hasofer, Design decision model considering mistake-proneness, in "*Risk, Structural Engineering, and Human Error*", edited by M. Grigoriu, Univ. Waterloo Press, pp. 163–189, 1984.

Firmage, D. A., *Modern Engineering Practice: Ethical, Professional and Legal Aspects*, Garland STPM Press, New York, 1980.

Fischoff, B., S. Lichtenstein, P. Slovic, S. L. Derby, and R. L. Keeney, *Acceptable Risk*, Cambridge Univ. Press, Cambridge, England, 1981.

Greenley, D. A., R. G. Walsh, and R. A. Young, *Economic Benefits of Improved Water Quality: Public Perceptions of Option and Preservation Values*, Westview Press, 1982.

Hachich, W. and E. H. Vanmarcke, Probabilistic updating of pore pressure fields, *J. Geot. Eng. Div., ASCE., 109*, 373–385, 1983.

Harford, J. D., Firm behaviour under imperfectly enforceable pollution standards and taxes, *Jour. Environmental Econ. and Mgmt.*, 5, 26–43, 1978.

Howe, C. W., *Natural Resource Economics: Issues, Analysis and Policy*, John Wiley, New York, 1979.

Kleindorfer, P. R. and H. Kunreuther, Insurance and compensation as policy instruments for hazardous waste management, *Intern. Conf. on Transport, Storage and Disposal of Hazardous Materials*, IASA, Laxenburg, Austria, July 1–5, 1985.

Kneese, A. V., and B. T. Bower, *Managing Water Quality*, Resources for the Future, Washington, D.C., 1968.

Lindley, D. V., *Making Decisions*, Wiley-Interscience, New York, 1971.

McDonald, J. F., The use of weighted discount rates in cost-benefit analysis: A further analysis, *Water Resour. Res., 17*, 478–480, 1981.

de Marsily, G., Spatial variability of properties of porous media:

A stochastic approach, in "*Fundamentals of Transport Phenomena in Porous Media*", edited by J. Bear and Y. Corapcioglu, NATO Advanced Science Institute, Series E, No. 82, pp. 719–769, 1984.

Massmann, J. and R. A. Freeze, Groundwater contamination from waste-management sites: The interaction between risk-based engineering design and regulatory policy. 1. Methodology, 2. Results, *Water Resources Research*, in press, 1987.

Neuman, S. P., Statistical characterization of aquifer heterogeneities: An overview, *Recent Trends in Hydrogeology, Geological Society of America paper 189*, 81–102, 1982.

Office of Technology Assessment, U.S. Congress, Protecting the nation's groundwater from contamination, *Report OTA-O-233*, Washington, D.C. 1984.

Paté-Cornell, M. E., Discounting in risk analysis: Capital vs human safety, in "*Risk, Structural Engineering and Human Error*", edited by M. Grigoriu, Univ. Waterloo Press, 17–32, 1984.

Raucher, R. L., The benefits and costs of policies related to groundwater contamination, Eastern Economics Association Annual Meeting, New York, N.Y., March 15, 1984.

Rothermal, T. W., Pitfalls in the hazardous-waste management business, *Environ. Prog.*, *2*, 221–225, 1983.

Routson, R. C., W. H. Price, D. J. Brown and K. R. Fecht, High-Level waste leakage from the 241-T-106 tank at Hanford, *Rockwell Hanford Operations, RHO-ST-14*, 1979.

Sharefkin, M., M. Schechter, and A. Kneese, Impacts, costs, and techniques for mitigation of contaminated groundwater: A review, *Water Resour. Res.*, *20*, 1771–1783, 1984.

Slovic, P., B. Fischoff, and S. Lichtenstein, S., Facts and fears: Understanding perceived risk, in "*Societal Risk Assessment: How Safe is Safe Enough?*" ed. by R. Schwing and W. A. Albers, Jr., Plenum Press, 1980.

Thurow, L. C., *The Zero-Sum Society*, Basic Books, 1980.

Vanmarcke, E. H. and H. Bohenblust, Risk-based decision analysis in dam safety, *MIT Research Report No. R82–11*, Cambridge, Mass., 1982.

Willard, D. E. and M. M. Swenson, Why not in your back yard? Scientific data and nonrational decisions about risk, *Environ. Mgmt.*, 8, 93–100, 1984.

R. ALLEN FREEZE
University of British Columbia,
Vancouver, B.C. V6T 2B4, Canada

The professional career of R. Allen Freeze began in 1961, shortly after he obtained the B.Sc. degree in Geological Engineering from Queen's University in Kingston, Ontario. In that year, he joined the Hydrologic Sciences Division of Canada Inland Waters Branch in Calgary, Alberta, as a Research Scientist. During the ensuing nine years, Al Freeze succeeded in earning the M.Sc. degree in Geological Engineering from the University of California, Berkeley (1964), and the Ph.D. in Civil Engineering from the same university (1966). During the last two years with the Hydrological Sciences Division, he was Head of the Western Research Section, Groundwater Subdivision.

From 1970 to 1973, Dr. Freeze was Research Staff Member in the Environmental Sciences Group of IBM at the Thomas J. Watson Research Center in Yorktown Heights, New York. In 1973 he joined the University of British Columbia as Associate Professor and Director of the Geological Engineering Program. Dr. Freeze became Professor in 1976, and from 1981 to 1984 he was Associate Dean in the Faculty of Graduate Students.

Dr. Freeze's major scientific and professional interest is in the solution of numerical and mathematical models of groundwater and surface water systems with the aid of computers. He applies the results to resource evaluation, rainfall-runoff modeling, and to the solution of geotechnical problems such as land subsidence, slope stability, seepage at dam sites, and waste-disposal-site analysis. He has authored over sixty technical and scientific publications in the fields of hydrology, hydrogeology, soil physics and engineering seepage. Allan Freeze co-authored with John A. Cherry in the classic text, "Groundwater."

In recognition of his many achievements, Dr. Freeze was honored by the American Geophysical Union with the Robert E. Horton Award in 1970 and in 1972 for the year's best paper in the field of hydrology; and with the James B. MacElwane Award in 1973 "in recognition of significant contributions to the geophysical sciences by a young scientist of outstanding ability." The

Geological Society of America conferred upon Allan Freeze the Meinzer Award in 1974, "in recognition of distinguished contributions to hydrogeology." In 1978 he received the Canadian Geotechnical Society Prize for "outstanding contributions to the geotechnical profession by a publication in the Canadian Geotechnical Journal." He was editor of *Water Resources Research* (1976–1980) and President-Elect and President of the Hydrology Section of the American Geophysical Union (1982–1986).

R. Allan Freeze is a Fellow of the Royal Society of Canada.

FOREWORD

Optimization of regional water resources systems has remained an elusive target since methods of decision analysis were welded onto models of hydrologic phenomena and economic factors. Continuous dissatisfaction with optimization was not for a dearth of models – quite the opposite, the plethora of mathematical formulations seemed to exceed by far the assimilative capacity of the profession. The dissatisfaction stems from some basic difficulties inherent in modeling regional water resources systems: the stochastic nature of hydrologic phenomena; and the strong socio-economic effects of water projects. The first issue was addressed in part in the Second Kisiel Memorial Lecture[1] and the second issue was alluded to in the Third[2] and also, to some extent, in the Sixth[3] The Seventh Kisiel Memorial Lecture addresses socio-economic issues in a direct manner.

Early optimization models of water resources development projects considered economic efficiency to be a necessary and sufficient criterion for the optimal design and operation of regional systems. Nevertheless a lingering doubt questioned the exclusive use of this criterion. After all, an efficient project increases the economic "pie" of a region and no one need be worse off than before. However, the distribution of the "pie" among its beneficiaries and those who paid for the project – matter of equity – was not resolved by the exclusive use of the economic efficiency criterion. A good illustration of this point is a study conducted in the state of Gujarat in India where high-yielding varieties of wheat and other crops were introduced ("the Green Revolution"). The study, which followed in detail over a period of more than ten years the performance of the improved agricultural production system, discovered that although all farmers in the area studied had higher

---

[1] Myron B. Fiering. The Real Benefits from Synthetic Flows. 1983.

[2] J.D. Bredehoeft, Water Management in the United States - A Democratic Process (Who are the Managers?), 1984.

[3] R. Allan Freeze, Groundwater Contamination: Technical Analysis and Social Decision Making, 1987.

net incomes, the range of incomes was greater than before the introduction of the new varieties. The greater disparity between incomes generated greater social tensions. Thus economic efficiency alone does not seem to be an adequate criterion for the optimization of regional water resources systems. Equity must be considered as a full partner among the criteria for the optimal design and operation of regional water resources development projects.

Economic efficiency and equity are not necessarily independent dimensions of a water system: there may be trade-offs between them. The Seventh Kisiel Memorial Lecture present a broad analytical panorama of this issue, offers answers to some problems, and raises additional stimulating questions which require, for their analysis and resolution, considerations transcending the more restrictive evaluations of water resources systems.

Nathan Buras

EFFICIENCY GAINS FROM BUILDING EQUITY INTO WATER DEVELOPMENT

CHARLES W. HOWE

I. INTRODUCTION

Economic efficiency and equity as objectives of water development have a long history in the evolution of water policy in the United States, in political and economic thought, and in the development of formal decision theory. Concern with these objectives is evidenced in Chester C. Kisiel's work, for example in "Interactive Multiobjective Programming in Water Resources: A Case Study" (with Monarchi and Duckstein, 1973). At least one prior speaker in this series, Peter Eagleson, noted the scope of regional (and perhaps global) hydrologic interdependencies and their implications for water planning (fourth memorial lecture, 1985):

> Because of humanity's sheer numbers and its increasing capacity to affect large regions, the hydrologic cycle is being altered on a global scale with consequences for the human life support systems that are often counterintuitive.

Among the examples of large regions of evaporative influence, Professor Eagleson cited the vast parts of Africa and the Atlantic Ocean influenced by evaporation from the Sudd, and great Nile swamp located in the Sudan. Those acquainted with the partially completed Jonglei Canal project to divert Nile River flow from the Sudd can infer from Eagleson's analysis that this one national project has the potential of influencing precipitation patterns throughout Africa. In these circumstances, it could well be that a project appearing socially desirable from a national viewpoint could be catastrophic from a continental or global viewpoint.

Is this an issue of economic efficiency or equity? It is clear that both are involved, that the most effective use of scarce water resources requires recognition of effects beyond national borders, as would most persons' concepts of equity. On the other hand, there are project or program situations in which economic efficiency and commonly held concepts of equity are in conflict.

This paper attempts to substantiate the following points: (1) that the situations in which economic efficiency and equity are in conflict are much rather than is commonly thought; (2) that water development in the United States, in particular, has been neither efficient nor equitable; (3) that many impacts of water development that tradition would have classified as equity impacts have, in fact, very important efficiency consequences; and (4) that an institutional (legal, administrative) setting that requires *actual* compensation to parties that sustain losses from projects (water and otherwise) will result in project selection and project designs that are more efficient than those likely to be chosen when full liability for costs is not present.

II. ECONOMIC EFFICIENCY AND EQUITY IN UNITED STATES WATER DEVELOPMENT

> "Economic decisions have a large zero-sum element. On average, society may be better off, but this average hides a large number of people who are better off and a large number of people who are much worse off. If you are among those who are worse off, the fact that someone else's income has risen by more than your income has fallen is of little comfort." (Lester C. Thurow, *The Zero-Sum Society*)

Little new evidence is needed to show that many of the post World War II water development projects in the United States have not been economically nor socially justified. The Warrior-Tombigbee Waterway (Carroll and Rao, 1978a) that cost nearly 2 billion dollars today attracts minuscule commercial traffic and accommodates mostly luxury pleasure craft – an outcome that was solidly forecast (Haveman, 1979) and that is certainly neither efficient nor equitable. It has been estimated that the present value of net losses of the project total 700 million dollars (Carroll and Rao, 1978b).

The famous snail darter case of Tellico Dam (Davis, 1988) called to the public's attention a public project that could not pass the test of comparing benefits to remaining costs after it was 95% complete. One of the significant cultural costs (whatever the outcome for the endangered snail darter) was the inundation of traditional tribal archeological sites as well as the inundation of the best bottom lands in the valley.

The Dolores Project is a Bureau of Reclamation irrigation project in Southwestern Colorado (Mann, Fisher and Young, 1986) that is fairly typical of recent federally sponsored irrigation. About 20,000 acres are to receive supplemental irrigation while 28,000 previously dry acres will receive supplemental irrigation service. Problems have arisen that have caused a large number of farmers who contracted for irrigation water to sue the Bureau to get out of their contracts. These contracts require repayment of the project's operating and maintenance costs which have escalated sharply since the project was planned. At the same time, real commodity prices have fallen, with the prospect of further decreases (Miller, *et al.*, 1986) in spite of recent drought conditions.

The Dolores Project was one of several Upper Colorado River Basin projects authorized by Congress in the Colorado River Project Act of 1968 as a *quid pro quo* for approval of the Central Arizona Project. Conditions for the project were never propitious. The project area lies at elevations exceeding 6,000 feet with a short growing season that varies from 112 to 140 days with frequent late spring and early fall frosts that preclude any crops other than forages, small grains, and edible dry beans. The terrain is sloping and broken, creating small field sizes and precluding surface irrigation. Side-roll sprinklers cost about $300 per acre, even though the federal project will deliver water already piped under pressure for sprinkler use. This extension of pressurized water delivery to the farm is part of the Bureau's "creative cost accounting" of recent times to place a larger share of costs under the irrigation subsidy for capital costs. It also has been used in the Animas-LaPlata Project in the same region of Southwestern Colorado.

The area is remote from markets and suffers already from surplus hay production from the Navajo Project to the south. As a result of these factors, Mann, Fisher and Young have concluded that the most efficient procedure would be not to finish the project. The fact that many farmers are suing to get out of their contracts indicates that this might be the most equitable path, too.

The Central Utah Project, which has yet to deliver any water, has been supported for 14 years by the Central Utah Water

Conservancy District that has taxed the residents of 12 Utah counties over that period. Some of these counties will receive water from the Project. The tax monies are being used as a war chest to promote acceptance of the project. In particular, the District has offered to build intake water treatment plants for towns that sign contracts for water (Bagley et al, 1983). As with other prominent Bureau projects (e.g. the Central Arizona Project), the Bureau proceeded without signed contracts and demand has been slow to develop for project water. In fact, the Bureau currently is buying agricultural water rights to substitute for the supplies that were to have been provided by the last phase of the project (Water Market Update, June 1987).

Equity has been touted as an objective of water development at least since the passage of the Reclamation Act of 1902, a major aim of which was to provide for the expansion of the family farm into the western lands. As has been well documented by Marc Reisner in the well researched and Popular *Cadillac Desert*, (1986), the effect of some prominent existing and proposed projects has been quite the opposite, with small producers being forced out of the market because of supplies expanded by very large producers (e.g. the Central Valley Project) or buying out family farms to construct reservoirs to serve large commercial operations (the proposed Narrows Dam on the South Platte River in Colorado). Howe and Easter (1971) long ago argued that much of western irrigation has simply displaced agricultural production in other parts of the country through the reduction of market prices based on highly subsidized supply costs of irrigation.

What has ruled western water development in fact is a criterion of "regional efficacy" and horse trading in which federal projects, the benefits of which accrue to the project region (often at substantial market costs to other regions), are largely paid for by the federal government in exchange for locating similar projects elsewhere (Ingram, 1972). Even newspaper advertisements have humorously and openly acknowledged the nature of these projects, including the case of the Animas-LaPlata Project in Southwestern Colorado. Large commercial landowners continue to be the primary beneficiaries, even (or especially after the Reclamation Reform Act of 1986 that increased the 160 acre acreage limitation to 960 acres, a limit that had been a standing joke for decades (see Dawdy, forthcoming).

III. THE RELATIONSHIP OF ECONOMIC EFFICIENCY AND EQUITY IN THEORY AND PRACTICE

Early post World War II treatments of public finance (i.e. the discipline treating the revenue and expenditure decisions of governments) frequently listed three objectives of governmental economic policy: (1) to allocate resources more efficiently; (2) to secure desirable adjustments in the distribution of income and wealth (equity); and (3) to stabilize the level of economic activity to avoid severe booms and busts. Richard A. Musgrave, [*Theory of Public Finance*], introduced an analytical framework that depicted government having three corresponding economic branches: (1) the allocation (efficiency) branch; (2) the distribution (equity) branch; and (3) the stabilization branch. Each of these branches had certain responsibilities that presumably could be carried out largely independently of the other branches.

While this approach had pedagogical and analytical advantages in dissecting the complex economic issues faced by governments at all levels, it created an attitude among economists that efficiency and equity could be considered separately – that if the most efficient program of public works, programs, and regulations had undesirable equity consequences, there would be income redistributional programs and social safety nets that would rectify the inequities. Everyone would then supposedly be better off than before the project. This conceptualization of governmental processes was found not only in courses in public finance but also in analyses of regional and Third World development. Economic development specialists from universities to the World Bank frequently asserted that economic development required investment embodying modern technologies and that these technologies could be expected to have distributional effects, particularly unemployment of part of the traditional labor force. The introduction of modern technologies was recommended nonetheless, as if inequities would be rectified by other means.

One certainly can find examples of trade-offs between efficiency and equity. All types of agriculture exhibit scale economies (decreasing unit costs of production) with respect to the area cultivated: peasant agriculture exhibits economies up to several acres; U.S. midwestern mixed farms exhibit decreasing unit costs up to about 400 acres or so; and dry land wheat exhibits economies up to several thousand acres. If farming operations are planned below these sizes to employ more farmers, there is a direct trade-off between efficiency and equity.

The "green revolution" that has done so much to increase the production of staple grains in the Third World has produced inequalities in wealth among farmers, since only those farmers with the capital and skills to provide the needed water, fertilizer, and pesticide inputs were in a position to adopt the new high-yielding varieties (Brown 1979). Thus, agricultural technology helps to define the available trade-offs between efficiency and equity.

Irrigation water supply systems exhibit interesting trade-offs, especially in a Third World context. Since the costs of delivering water is a function of the distance from the source (because of canal costs and water losses), efficiency calls for delivering less water to more distant lands than to close-by lands during periods of water shortage. (This efficiency condition must not be confused with the frequently observed fact that the rich and powerful tend to own the lands at the head of the system, often taking much more water than warranted by efficiency guidelines while leaving far too little for the unfortunates at the foot of the system. See Wade, 1984.) Many societies feel that such a distribution of water is inequitable and operate their systems to provide roughly the same water per acre to all lands, regardless of distance. The PASTEN system in Indonesia is of this type (see Howe, forthcoming). As water becomes increasingly scarce, the trade-off between efficiency and equity becomes sharper under the PASTAN system. (See Carruthers, 1983); Easter, 1986; Taylor and Wickham, 1979).

Regarding the theory of efficiency-equity trade-offs, benefit-cost analysis has dealt with equity in two major ways: (1) through constraints on project location, design, and operation; and (2) by assigning weights to benefits and costs, depending on whom they fall. Examples of the first approach would be investment approval rules requiring that the benefit cost ratio exceed 1.0 or some higher number, while also requiring that projects provide some net benefits for the poorest segments of population, avoid destruction of cultural assets, maintain the employment of women, etc. Naturally, each of these additional requirements may reduce the net economic efficiency benefits.

The second approach has been advocated by a number of influential writers including Squire and van der Tak (1975), and Ray (1984) of the World Bank, to weight project benefits and costs with "social weights" to reflect the importance of benefits accruing to various groups and of costs being extracted from different groups. Thus, if a country shares an egalitarian philosophy, a weight of 12 might be assigned to benefits accruing to the poor, 1.0 for middle class benefits, and 0.9 for benefits to the rich. With this approach, efficiency and equity are blended into one measure of project desirability.

In the benefit-cost analysis of projects with uncertain (or risky) payoffs, the theoretically correct measures of efficiency and equity effects also blend into one measure of project efficacy (see Graham, 1981). Consider an irrigation project that will produce net benefits for a typical farmer of $50 during a wet year and $100 during a dry year. Suppose that wet years occur with probability 0.7 and dry years with probability 0.3. The expected (or average) payoff would then be $65 for the typical farmer. Most benefit-cost analyses would pragmatically use this value as the annual benefits per farmer. However, $65 is unlikely to reflect the real value of the project as perceived by different farmers. Some farmers are well-to-do from non-farm sources, others are poor. Some are highly risk-averse, others may be risk-seeking. A detailed, idealized benefit-cost analysis would consider equivalent to the 70%, 30% probabilities of payoffs of $50 and $100 (see Graham, 1981). The aggregated value of these option prices then is the appropriate annual benefit measure, but it clearly depends on just *which* farmers are recipients of the probabilistic benefits, i.e. it cannot be calculated without knowing who receives benefits and who pays the costs.

Thus efficiency-equity trade-offs exist in the real world in many situations. Project evaluations can present either the standard present value of net benefits measure, supplemented by data on incidence on benefits and costs, or can present the present value of socially weighted benefits and costs or aggregated option prices as a single index of project desirability. In this author's experience, most decision makers would prefer the former detailed data presentation to the latter summary data.

In actual water planning practice, equity considerations get little attention at all, partly because those negatively affected by a project frequently are poor and have no political voice and partly because planners fail to realize that "equity" dimensions of project impacts often have important efficiency consequences. In the case of the Kousou Dam on the Bandama River in Ivory Coast, the population displaced by the reservoir was considered sufficiently unimportant that resettlement warranted one page in the project feasibility report: give each person some petty cash and they will resettle themselves. After extended tribal warfare and after completion of a $400 million resettlement program (more than the cost of the dam), the government and the project sponsors came to realize that trying to ignore equity issues could be a costly mistake, negating any net benefits that might have been generated by the project.

In many water project feasibility studies, the boundaries of the "project" are either naively or intentionally drawn too small to encompass all relevant project effects. This practice can result in both inefficiencies (by ignoring external costs) and inequities (by ignoring negative impacts on disadvantaged groups). In the case of the Kainji Dam on the Niger River in Nigeria, the project was justified on the basis of new irrigation and hydropower development. No consideration was given to the impact of river regulation on downstream recession agriculture, livestock systems, and fisheries. It is now estimated that the economic losses to those systems exceed the net value produced by the newly irrigated areas and hydropower (Scudder, 1989, Ch. 3).

Projects like the Animas-LaPlata in Southwestern Colorado ignore many of the external costs imposed on other parties. In the Animal-LaPlata case, downstream reduction in hydroelectric generation, reductions in irrigated output and increases in salinity amount to at least $99 per acre-foot of additional consumptive use (Howe & Ahrens, 1988), yet have not been counted as project costs.

Similarly inadequate analyses accompany most out-of-basin water transfer projects that impose a long, complex line of indirect damages on the basin-of-origin (MacDonnel & Howe, 1986). Stream flows are reduced, recreational and aesthetic values diminished, water quality is reduced and related costs increased, activities dependent on recreation are reduced, and future development is sometimes precluded. In most western states, the diverting party is liable for few if any of these indirectly imposed costs, thus motivating projects that may be economically inefficient overall and highly inequitable to the basin-of-origin.

In summary, in industrialized and developing countries alike, there is no other branch of government that will rectify inequities caused by projects. Equity has to be built in at the project level or it never will be accomplished. While there are genuine equity-efficiency trade-offs in some cases, many "equity" impacts that planners prefer to ignore turn out to have strong long-run efficiency consequences, such as the failure of local farmers to maintain irrigation systems that have been forced upon them (see Uphoff, 1986).

IV. BETTER WATER UTILIZATION FROM COMPULSORY COMPENSATION OF LOSERS AND IMAGINATIVE BENEFIT SHARING

Economists have used several efficiency concepts over time. In the "new welfare economics" of the 1930's to 1950's, economists tried to eschew value judgements by applying the concept of "Pareto optionality" to the assessment of projects or program: a project is desirable (efficient) if no one is worse off as a result of the project and at least one person (hopefully many) is better off than before (see e.g. Hirshleifer, 1984, Ch. 15). Very few real world projects can be judged by this criterion since some persons or firms are left worse off by any project. In many cases, a redistribution of benefits (e.g. by cash payment, provision of services, etc.) could avoid leaving others worse off, but the mechanisms for redistribution are usually not present.

Rather than being left unable to judge any project's economic desirability, the British economists Nichols Kaldor (actually Hungarian) and John Hicks suggested the "compensation principle" for evaluating projects and programs: if the winners from a project could fully compensate all losers from that project out of their benefits and still be better off, the project should be declared economically desirable, whether or not compensation actually takes place. As just stated, the principle does not avoid value judgements since, in the absence of actual compensation, the principle implies that the social utility of benefits accruing to the winners is greater than the social disutility imposed on the losers. Two different institutional frameworks are being compared: one in which compensation is mandatory and one in which it does not take place.

Benefit-cost analysis is based somewhat loosely on the compensation principle, asserting that if aggregate benefits (gains to the winners) exceed aggregate costs (losses to losers), the project passes the test. In practice, the distribution of benefits and costs is ignored. The Flood Control Act of 1936 required that benefits exceed costs "to whomsoever they accrue," in the same spirit. Market prices are presumed to reflect marginal benefits to winners and marginal costs to losers. Naturally, a given project may affect winners and losers in very non-marginal ways, making the measurement of benefits and costs considerably more difficult (i.e. consumer and producer surpluses must be included).

What is the appropriate measure of compensation to a party that loses? As an example, consider the family whose farm is taken by the project for the reservoir area. If the family were just indifferent between staying and selling the farm at the existing, market price, this price would constitute adequate compensation. Typically, however, these farm families are *not* indifferent and prefer to stay where they are: the market price is *not* an adequate measure of their "willingness-to-accept" compensation for the farm, else they would have sold and left. There is continually growing evidence that people have particularly strong feelings about losses, that their willingness-to-pay to gain an asset is less than their willingness-to-accept for giving up the same asset--especially if the asset is important or monetarily large (see Cummings, Brookshire, & Schulze, 1986, Ch. 3). Courts will assign the market value (say, of similar properties) as adequate compensation if condemnation procedures are used. The Bureau of Reclamation even used "block busting" tactics to induce panic selling for the Narrows Project on the South Platte River (Reisner, Ch. 11).

Law, both case law and legislation, define the extent of liability, of times with little relationship to economic realities. Early appropriations doctrine (and still found largely unchanged in Colorado) made water diverters liable for injury to other diverters, but with no mention of instream, aesthetic, or water quality values. Colorado law requires water conservancy districts (but not cities) to provide compensatory storage for transfers out of the Upper

Colorado Basin, a very uneven and generally inefficient form of compensation. Colorado has made it possible to dedicate water rights to instream flow maintenance, but there is no protection of instream values when water is initially appropriated or subsequently transferred to new uses. Water quality is not protected except through occasional stipulations in water court settlements (see Getches, 1984).

Many western states have developed sets of social criteria that must be met prior to the authorization of new appropriations or transfers. State laws have identified public values other than those accruing to water diverters that are to be protected in the administration of water: water quality, instream flow values, local economic base, and the family farm are found as criteria in Idaho, Utah, and Wyoming. New Mexico has established the public welfare of the people of New Mexico and appropriate conservation of water as criteria--whatever they may mean. The effect of these changes is to force recognition of a broader set of values and to increase liability for damages, i.e. to force internalization of an increasing range of externalities.

Interstate compacts rule the allocation of many western interstate streams (e.g. the Colorado, the Rio Grande, the Pecos, etc.) in ways that were perceived to be equitable if not efficient at the time of compact negotiation. Over time, some of these compacts have become outdated--out of touch with development in the riparian states. Yet the allocations continue to be made with *no* comparison of benefits and costs. Such compacts, in the absence of more imaginative compensation schemes, stimulate the development of an "it's our water, develop it at any cost" syndrome.

In a lecture in Durango, Colorado, in November, 1987, I tried hard to develop and present analyses that showed the Animas-LaPlata Project to be against not only national economic and environmental interests, but also against state and possibly even local interests-the latter in spite of the enormous subsidies alluded to in Figure 1. At the end of the lecture, a well-known and thoughtful citizen of the community asked the question, "Given that the project is terribly inefficient, where else are we going to get development money for the community?" This was a very legitimate question, for Southwestern Colorado is a chronically depressed region, while federal regional development funds have disappeared.

Another participant voiced local feelings more colorfully, "You know that $350 million B-1 bomber that crashed in Eastern Colorado last month? At least we'll have something to show for our $350,000 project." The community's unique hope was that the construction period would produce some jobs and sales, whatever the completed project might do. Five percent of the construction cost given to the city for infrastructure, schools, and promotion of the region for recreation would have yielded far greater benefits than the project itself.

The sharing of benefits has long been recognized as necessary for the optimization of multi-jurisdictional projects. Jones, Pearse, & Scott (1980) set forth the necessary conditions of benefit sharing to prevent the sub-optimal, piecemeal development of a resource like a river. The bi-national development of storage and generating capacity of the Columbia River is a good example of benefit sharing with cash payments going from the United States to Canada to compensate for the development of superior storage in Canada while power was generated in the United States where the demand was (see Krutilla, 1967).

Practical suggestions have been made for multi-state benefit-sharing that would produce an environment in which more national water development could take place. Richard Lamm, then Governor of Colorado, suggested that sharing part of the electric power revenues from the federal installations on the Colorado River with the State of Colorado would create motivation for Colorado not to push so hard for further inefficient, water-consuming projects in the State. The question of the redistribution of those funds to parties who would benefit from (subsidized) water projects remains, but the public at large would appreciate that water left in the stream has value.

In closing, we recognize that compensating damaged parties and benefit-sharing are different activities, but that both help create an environment in which full system costs and benefits will be recognized. The United States' water establishment, federal and otherwise, has been very unimaginative and uncaring when it comes to equity and broad systems optimization. Policies that require compensation of a broad range of damages and that facilitate imaginative benefit sharing will greatly improve both the efficiency and the equity of water development.

LITERATURE CITED

Bagley, Jay M., Frank W. Haws, Wendell B. Anderson, Daniel H. Hoggan, & Lee Kapaloski, 1983, "*Impediments to Effective Interactions Between Multipurpose Water Districts and Other Governmental Institutions in Urbanizing Areas*,"Project No. B-170-UTAH, Utah Water Research Laboratory, Utah State University, Logan, UT, July.

Brown, Lester R., 1979, *Seeds of Change: The Green Revolution and Development in the 1970's*, New York, Praeger Publishers.

Carroll, Joseph L. & Srikanth Rao, 1978a, "The Role of Sensitivity Analysis in Inland Water Transport Planning: The Case of the Tennessee-Tombigbee,"*Proceedings of the Nineteenth Annual Meeting of the Transportation Research Forum*, Volume XIX, No. 1, pp. 46-54.

Carroll, Joseph L., & Srikanth Rao, 1978b, "Economics of Public Investment in Inland Navigation: Unanswered Questions," *Transportation Journal*, Spring, pp. 27-54.

Carruthers, Ian D., 1983, *Aid for the Development of Irrigation*, Paris: Organization for Economic Cooperation and Development.

Cummings, R.G., D.S. Brookshire, and W.D. Schulze, 1986, *Valuing Environmental Goods: An Assessment of the Contingent Valuation Method*, Totowa, N.J.: Rowman & Allanheld.

Davis, Robert K., 1988, "Lessons in Politics and Economics from the Snail Darter" in V. Kerry Smith (ed.) *Environmental Resources and Applied Welfare Economics*, Washington, D.C.: Resources for the Future, Inc.

Dawdy, Doris Ostrander, forthcoming 1989, *Congress in Its Wisdom: The Bureau of Reclamation and the Public Interest*, Boulder, CO: Westview Press.

Eagleson, Peter S., 1985, "The Emergence of Global-Scale Hydrology," Fourth Chester C. Kisiel memorial lecture, Department of Hydrology & Water Resources, University of Arizona, Tucson, March 20th.

Easter, K. William (ed.), 1986, *Irrigation Investment, Technology, and Management Strategies for Development*, Boulder, CO: Westview Press.

Getches, David H., 1984, *Water Law in a Nutshell*, St. Paul, MN: West Publishing Co.

Graham, Daniel A., 1981, "Cost-Benefit Analysis Under Uncertainty," *American Economic Review* 71(4): 715-725, September.

Haveman, Robert H. 1979, Affidavit presented to the United States District Court for the Northern District of Mississippi in connecting with the Tennessee-Tombigbee Waterway Project.

Hirshleifer, Jack, 1984, *Price Theory and Applications* (Third Edition), Englewood Cliffs, N.J.: Prentice Hall, Inc.

Howe, Charles W. & K. William Easter, 1971, *Interbasin Transfers of Water: Economic Issues and Impacts*, Baltimore and London: The Johns Hopkins Press for Resources for the Future, Inc., Chapter 6.

Howe, Charles W., forthcoming, "Equity versus Efficiency in Indonesian Irrigation: An Economic Evaluation of the PASTEN Method," Chapter 9 in R.K. Sampath and Robert A. Young (eds.), *Socio-Economic Issues in Irrigation Management*, Boulder, CO: Westview Press.

Howe, Charles W. & W. Ashley Ahrens, 1988, "Water Resources of the Upper Colorado River Basin: Problems and Policy Alternatives," Chapter 5 in Mohamed T. El-Ashry and Diana C. Gibbons (eds.) *Water and Arid Lands of the Western United States*, New York: Cambridge University Press.

Ingram, Helen, 1972, "The Changing Decision Rules in the Politics of Water Development,"*Water Resources Bulletin*, Vol. 8, No. 6, December.

Jones, R.A., P.H. Pearse, and A.D. Scott, 1980, "Conditions for Cooperation on Joint Projects by Independent Jurisdictions," Canadian Journal of Economics, May.

Kisiel, Chester C., David Monarchi, & Lucien Duckstein, 1973, "Interactive Multiobjective Programming in Water Resources: A Case Study," *Water Resources Research*.

Krutilla, John B., 1967, *The Columbia River Treaty: The Economics of An International River Basin Development*, Baltimore: The Johns Hopkins Press for Resources for the Future, Inc.

MacDonnell, Lawrence J. & Charles W. Howe, 1986,"Area-of-Origin Protection in Transbasin Water Diversion: An Evaluation of Alternative Approaches,"*University of Colorado Law Review*, 57(3): 527-548, Spring.

Mann, Roger, Adrian G. Fisher and Robert A. Young, 1986, "Economic Aspects of Utilizing Dolores Project Irrigation Water in Southwest Colorado," Department of Agricultural and Natural Resources Economics, Colorado State University, Fort Collins, Co, April.

Meyers, Charles J., 1966, "The Colorado River,"*Standing Law Review*, Vol. 19, No. 1, November, pp. 1-75.

Miller, T.A., W. Track & D. Smith, 1986, "Future Prospects for Agriculture in Colorado and the United States," Bulletin 533A, Colorado State University Extension Service, Fort Collins, CO.

Musgrave, Richard A., 1959, *The Theory of Public Finance: A Study In Public Economy*, New York: McGraw-Hill Book Co.

Ray, Anandarup, 1984, *Cost-Benefit Analysis: Issues and Methodologies*, Baltimore & London: The Johns Hopkins University Press for the World Bank.

Reisner, Marc, 1986, Cadillac Desert: *The American West and Its Disappearing Water*, New York: Viking Penguin, Inc.

Scudder, Thayer, 1989, *The African Experience with River Basin Development: Achievements to Date, The Role of Institutions, and Strategies for the Future*, Boulder, CO: Westview Press.

Squire, Lyn and Herman G. van der Tak, 1975, *Economic Analysis of Projects*, Baltimore & London: The Johns Hopkins University Press for the World Bank.

Taylor, Donald C. & Thomas H. Wickham, 1979, *Irrigation Policy and the Management of Irrigation Systems in Southeast Asia*, Bangkok: Agricultural Development Council.

Thurow, Lester C., 1980, *The Zero-Sum Society: Distribution and the Possibilities for Economic Change*, New York: Basic Books.

Uphoff, Norman, 1986, *Improving International Irrigation Management with Farmer Participation: Getting the Process Right*, Boulder, CO: Westview Press.

Wade, Robert, 1984, "Irrigation Reform in Conditions of Populist Anarchy," *Journal of Development Economics*, 14(3): 285-303.

Water Market Update, June 1987, Vol. 1, No. 6, p. 2. Shupe and Associates, P.O. Box 2430, Santa Fe, NM 87504.

White, Gilbert F. (ed.), 1977, *Environmental Effects of Complex River Development*, Boulder, CO: Westview Press.

CHARLES W. HOWE
University of Colorado
Boulder, CO 80309
U.S.A.

Chuck Howe's commitment to the study of water resources began when, as a transportation economist and operations researcher at Purdue University, he undertook studies of the inland waterway system of the United States. These studies, sponsored by the Ford Foundation and Resources for the Future, Inc. resulted in an article "Methods for Equipment Selection and Benefit Evaluation in Inland Waterway Transport" that appeared in the first issue (Vol. 1, No. 1, 1965) of Water Resources Research and a book, Inland Waterway Transportation, published by Resources for the Future in 1969. In the course of this research, Howe became acquainted with and greatly influenced by three of the great figures in the physical and policy sciences of water: Walter V. Langbein, Allen V. Kneese, and John V. Krutilla. Langbein and Kneese were the first co-editors of Water Resources Research.

After joining Resources for the Future as Director of the Water Resources Program, Howe followed Kneese as Policy and Social Sciences Editor of WRR, serving from 1967 through 1974. During the RFF years, Howe collaborated with Pierce Linaweaver, Jr. on an early study of the influence of price and other socio-economic variables on residential water demand, published as "The Impact of Price on Residential Water Demand and Its Relation to System Design and Price Structure" (WRR, Vol. 3, No. 1, 1967). With co-author K. William Easter, he wrote the book Interbasin Transfers of Water: Economic Issues and Impacts (Resources for the Future, 1971). With the help of Langbein, he arranged for the innovative work on the economics of groundwater by John Bredehoeft and Robert Young.

Questions of social equity in the development of water resources were first raised during Howe's work in Third World countries: Ghana, Mexico, Botswana, Kenya, Indonesia, and, most recently, Gambia and Senegal. It became clear that water systems, especially irrigation systems, were often designed to make the rich richer. If equity is not considered in the design of water systems and their operating rules, it will not be achieved by other means. If equity is explicitly considered, it will frequently result in greater economic efficiency of the system--not less-- and in greater environmental sustainability. These points are developed in this lecture.

Chuck Howe is a Fellow of the American Geophysical Union and serves as President of the Association of Environmental and Resource Economists.

FOREWORD

At a meeting in Haifa, Israel, in March 1967, I asked the late Professor Léon J. Tison, the Belgian scientist who served for many years as the secretary of the International Association of Scientific Hydrology (later named as the International Association of Hydrological Sciences), what was the most important problem to be studied in hydrology. His immediate, clear and unequivocal answer was: the vadose zone. Little did I suspect at that time that twenty-two years later I shall have the privilege of introducing one of the most active, prolific, and forward-looking scholars of the flow phenomena in the unsaturated porous media.

Important physical, chemical and biological processes, not completely understood, occur within the vadose zone. As a consequence, the unsaturated layer below the land surface acquires a special importance, both from a purely scientific point of view and for practical reasons. Scientifically, the vadose zone is a critical link in the hydrological cycle in nature, controlling the rate at which part of the precipitation will percolate through soils to join, eventually, the groundwater stored in aquifers. Operationally, the vadose zone is directly involved in managing water quality, primarily groundwater. For example, the flow of water in the unsaturated zone of soils in semi-arid climates is dominated by the amount falling on the surface and reduced by evapotranspiration, as contrasted with temperate climates where this flow is dominated by the horizontal, saturated movement. As a consequence, contaminants discharged into the environment may be diluted while transported by water percolating through the vadose zone in the Eastern United States, or concentrated by the same process in the Southwest.

For many decades, the development of water resources was equated with engineering projects that meet a given (and/or projected) demand for water at minimum costs. More recently, with the increased competition for finite and limited water resources, we began to question the sanctity of the given demand and realized that the development of water resources requires an equilibrium between water availability, demand for water, and costs involved in meeting the demand. We realize now that the boundaries of the system need to be further extended to include a consideration of the resource after its use and with the attendant improvement in its quality.

The eighth Kisiel Memorial Lecture focuses on one of the most critical elements in any scheme for the management of water quality.

Nathan Buras

A CHALLENGING FRONTIER IN HYDROLOGY - THE VADOSE ZONE

DONALD R. NIELSEN

INTRODUCTION

Thirty-five years ago this spring, with an excellent educational experience here at the University of Arizona, I left Tucson for Iowa State College to continue my education studying soil physics. At that time no hydrology program existed at U. of A. And even if it had, I doubt that it would have been of interest to me. My roots stemmed from irrigated agriculture in the Salt River Valley and I envisioned a career in soil chemistry. The late Professor Theophil F. Buehrer, Professor and Head of the Department of Agricultural Chemistry and Soils, served as my major professor. And even before I completed the B.S. degree, Professor Wallace H. Fuller extended my vision to include microbiology. His leadership in the use of radioisotopes gave me the opportunity to trace the environmental fate of phosphorous fertilizers using the then brand-new isotopic dilution method. And through his encouragement, I proceeded into the physical sciences at Iowa State arriving shortly after the departure of the yet-to-be U. of A. Professor Daniel D. Evans. Upon my graduation, the University of California, Davis, expected me to conduct research on how to better irrigate vegetables for maximum production -- a task I have yet to undertake! The yet-to-be U. of A. Professor Gordon R. Dutt was at Davis developing a chromatographic model to describe the leaching of soluble salts in irrigated fields, and the future U. of A. Professor Peter J. Wierenga and currently Chair of the Department of Soil and Water Science was studying thermal regimes of rice fields irrigated with frigid water. Dr. L.G. Wilson was, at that time, finishing the Ph.D. degree at Davis after observing water and solute migration in diatomaceous earth in Northern California. U. of A. Professor Arthur W. Warrick, whom I first met at Iowa State University, continues to contribute to my understanding of the physics and geostatistics of the vadose zone inasmuch as today we are both members of a regional research project. Dr. Roy Rauschkolb, former U. of A. Director of Cooperative Extension and now Director of the Maricopa Agriculture Center, enlightened my perceptions of the placement and timing of liquid fertilizers in drip irrigation as regards their crop availability and decreased pollution potential. Professors Soroosh

Sorooshian and Shlomo P. Neuman have been my colleagues in editing the Water Resources Research journal. Professor Evans continues to remind me of the importance of water flow in fractured rock. All of these University of Arizona colleagues as well as others have had a bearing on my activities to explore the vadose zone. Some of their help has been by design while others have been by chance -- a characteristic of hydrology! These introductory remarks manifest interpersonal relationships that are both rewarding and significant in science but also signal the diverse, unstructured inputs that left me to begin to explore the vadose zone. The vadose zone remains the least explored frontier of hydrology.

The vadose zone is the water unsaturated zone of the earth's crust, being that land region bounded at its top by the soil surface and below by water-saturated geologic formations or by the groundwater table. The name "vadose" stems from the Latin noun vadosus meaning "shallow." In the humid regions of eastern U.S., its depth is indeed shallow, and hardly more than the root zone of cultivated crops or forested areas. While in western U.S., its depth can be of tens or hundreds of meters. Because of its thickness in the west as well as limited rainfall, water below the root zone takes a relatively long time moving vertically before traveling horizontally within the saturated water table. In the east, water movement below the soil surface is dominated by saturated, horizontal movement. In the west, the rate at which water moves through the vadose is controlled by the amount of water falling on the soil surface discounted by that transpired by plants and lost by evaporation and runoff at the soil surface. For example, in the Salt River Valley without irrigation, it would take decades for water to travel 10 meters below the soil surface. On the other hand, in the east where precipitation is more abundant, the hydraulic properties of the soil dominate the routing of water between runoff and percolation through the soil profile. And in the east, within only a year, water travels through the vadose zone on it way through saturated, more horizontal pathways. In the east excessive rainfall dilutes contamination, while in the west contaminants are concentrated. Thus, the vadose zone is the conduit through which water and its constituents are attenuated, concentrated, and transformed as they travel between the soil surface and the water table.

The recognition of the vadose zone amongst hydrologists today stems from the need to improve and protect the quality of groundwater supplies. The migration of fertilizers and pesticides from agricultural and domestic usage, of solvents and toxic substances from industrial usage, and of countless other inorganic and organic chemicals into the topsoil and through the unsaturated zone has signalled the pollution of groundwater. It has also been recognized for its potential in arid regions to store, retain, and confine unwanted waste materials from spreading and contaminating our global environment. As a result, state and federal legislation has initiated measures to control or regulate the kinds of chemicals being released directly or indirectly into the soil surface and to identify and delineate local environmental conditions that mitigate against chemical transport.

Although the motivation to understand and exploit the vadose zone has never been higher than it is today, ambiguities persist regarding its conservation and management. It has not been the consistent focus of attention of any scientific discipline. It has been ignored in most educational programs of higher education. The University of Arizona is the singular exception inasmuch as it has research and education programs that treat the vadose zone. The existence of the zone is not even perceived by the citizenry of our country. Until now there has been no incentive to include it in a scheme of natural resource management. It is there, but nobody cares!

Biological research for agriculture and silviculture considers its top boundary primarily as a cyclic source of water and plant-essential nutrients. The focal point of that research is the transformation of solar energy and the absorption of inorganic constituents to enhance the biotic potential of a region. Crop yield is foremost, with the alteration of the underlying vadose environment seldom considered. Hydrologists view it traditionally as a buffer for runoff and erosion through its potential to absorb water infiltrating from rainfall, or as a source of water that reaches the water table. Geochemists have generally ignored contemporary transfer events in favor of paleontological studies. And, microbiologists, recognizing that readily available carbon sources are more abundant in the topsoil in the vicinity of plant roots, have not been inclined to study the nature of microbial communities in the vadose zone. Hence, segmented interest or apathy has contributed to our lack of theoretical and experimental understanding of the vadose zone without a technology to adequately predict or manage its behavior.

The challenges for both fundamental and applied research to reveal the intricacies of the zone await those having an education and curiosity of sufficient magnitude to respond. Attempts to date have been fragmented between several disciplines without a unifying approach to bring together the pertinent physical, chemical, and biological features of the zone.

EARLY PIONEERS

Most of us get started on the wrong foot as regards our education about the unsaturated zone. Introductory college courses in the agricultural and earth sciences still profess the misleading concept of field capacity -- the property of a soil to be able to fill with water up to a threshold value before it begins to leak and transmit water to greater depths. Similarly, in engineering courses the concept of specific retention is described in an analogous manner to field capacity and leads to the companion term specific yield -- the water that drains to greater soil depths as a result of the specific retention or field capacity being exceeded. The concept is simple, readily understood, and facilitates a convenient set of numerical exercises given in the classroom as homework problems. Its origin and utility stem from a USDA bulletin published nearly a century ago (Briggs, 1897). In the context of those times, the concept was a step forward. Scientists were then grappling with ideas to understand the balance between soil water retention and its extraction by plant roots. As it has turned out, that concept diverted attention from the true nature of the unsaturated zone persisting even today. Unfortunately, real soils don't obey the concept. In real soils, water with its dissolved constituents can move through a soil layer without changing its water content. In other words, increases or decreases of soil water content at a particular soil depth are not sufficient to ascertain the direction and magnitude of the rate at which water flows through an unsaturated soil.

An understanding of the dynamics of how water moves through the vadose zone begins with the pioneering work of three persons -- three heroes of the unsaturated zone -- Edgar Buckingham, Willard Gardner, and Lorenzo A. Richards. Edgar Buckingham, born in Philadelphia in 1867, responsible for the Buckingham Pi theorem of dimensional analysis, and a brilliant physicist who authored the book "Outline of the Theory of Thermodynamics" in 1900, worked form 1902-1905 in the U.S.

Department of Agriculture. During that brief three-year period, he suggested that a relation exists between the soil water content and the energy status of the soil water analogous to the theories of thermal and electrical potentials. In addition, he suggested that the rate at which water moves through a soil is analogous to the manner in which heat and electricity are described by Fourier's and Ohm's laws. He proposed for soil that

$$\text{water flow rate} = -K\nabla\psi$$

where ψ is a quantity that measures the attraction of the unsaturated soil for water, and K the "capillary conductivity." His original description (1907) was profound, lucid and a challenge:

> "The analogy, however, is only formal. In the first place, the thermal and electrical conductivities of a given piece of material are independent of the strength of the current and, in general, only slightly dependent on the temperature and other outside circumstances, so that for most purposes they may be treated as constants. The capillary conductivity, however, we have every reason to expect to be largely dependent on the water content of the soil, and therefore variable, not only from point to point in the soil, but also with the time at any given point. For it is not to be expected that the ease with which water flows through the soil will be independent of the extent and thickness of the water films through which, i.e., along which, it has to flow.
>
> "Furthermore, the other factor in the equation, namely, the gradient (ψ) is not the space variation of a simple and directly measurable quantity like a head of water, an electrical potential, or a temperature. It is the gradient of a quantity ψ, the attraction of the soil for water; and depends in some as yet unknown way, differing from soil to soil, on the water content of the soil, which can itself be measured only by tedious and not very accurate methods."

Willard Gardner, following the footsteps of Buckingham, led the development of a simple, easy method of measuring the capillary potential. Buckingham used vertical soil columns draining in the laboratory to merely infer its value from the assumption that the force of gravity was balanced by the gradient of the capillary potential when water ceased to drain. In 1922, Gardner placed a water-filled porous cup into the unsaturated soil. By observing the level of water in a manometer connected to the porous cup, he interpreted the value of the negative pressure within the cup to be a measure of the capillary potential. This simple technique prompted others to develop the well-known tensiometer used commonly today in conjunction with electrical pressure transducers to automatically record the energy status of water in vadose zone. Gardner, a physicist, prompted many young scientists at Utah State University to seek careers studying the vadose zone. One of them was Lorenzo A. Richards. In addition to his extensive work on instrumentation for quantifying soil water properties, he laid the foundation for water transport in unsaturated soils in 1931 with the derivation of an equation which is now known as the Richards' equation. The terms in that equation and its solution remain the primary focal point of contemporary research. Even with today's computer technology, solutions of the equation are scarce and their application to real world practical problems remain at nearly the same point we were at in 1931 when Richards completed his pioneering Ph.D. dissertation at Cornell University. Let's review some of the properties of the vadose zone, comprised of soils and sedimentary deposits, that have slowed our progress or remain a challenge to quantify.

INTERACTIONS WITHIN THE VADOSE ZONE

Soil is unique amongst all materials in the universe as regards its transmission properties. Alas, this was the major concern of Buckingham because he clearly recognized that the analogy of soil water properties to those of electricity and heat were imperfect. For the latter, electrical and thermal properties are practically independent of the amount of current or heat being transmitted, whereas soil water properties depend strongly upon the soil water content. Figure 1 shows this soil water content dependent behavior of the hydraulic conductivity. As the water content decreases from 42 to 36%, the hydraulic conductivity decreases 10-fold. Indeed, the hydraulic conductivity decreases 10 million times as the soil dries from water saturation to complete dryness. This uniqueness is a practical benefit in that it accounts for a soil's ability to rapidly absorb and retain rainfall under a variety of initial and local conditions. On the other hand, it increases the difficulty of obtaining solutions of Richards' equation as well as markedly exacerbates our ability to deal with the naturally occurring variations of soil properties across the landscape.

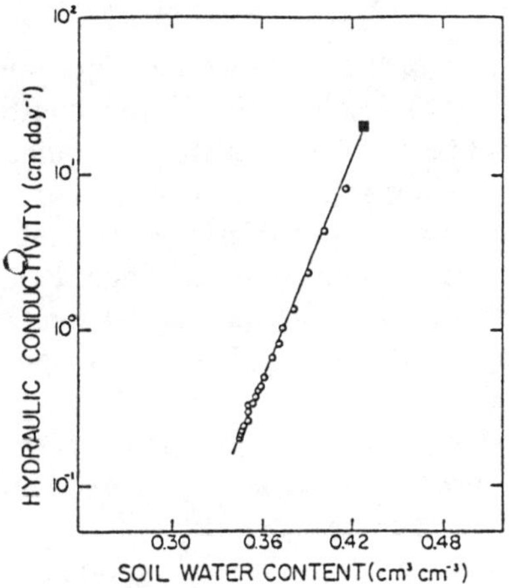

Figure 1. Measured values of soil hydraulic conductivity versus soil water content.

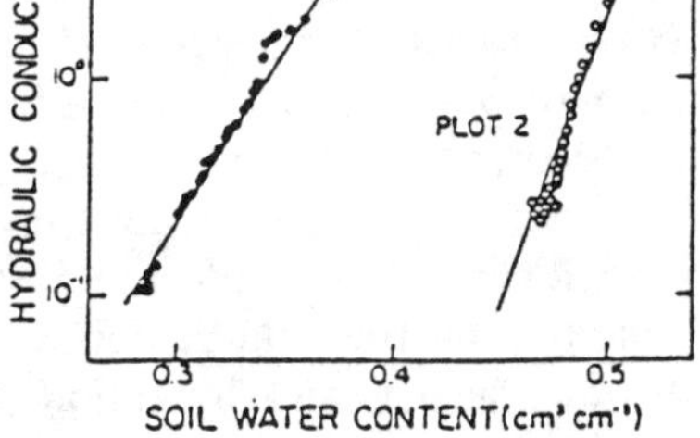

Figure 2. Hydraulic conductivity as a function of soil water content measured at the same soil depth in two plots separated by 50 m.

Figure 2 shows where the unsaturated hydraulic conductivity measured at two different locations separated by a distance of only 50 m is plotted against soil water content. Random variations of field-measured soil water content typically range $\pm$ 0.05 cm^3/cm^3. Note that the two curves differ by nearly two orders of magnitude for the same water content. Such variability forces estimates of the hydraulic conductivity even at short distances to have an uncertainty of 100-fold!

The water in the vadose zone is not pure, but is a solution of water and dissolved solid and gaseous constituents. Moreover, soil water cannot be considered simply as ordinary water with a few dissolved solutes, since soil water properties are intimately linked to the chemical and physical properties of the solid phase on which it is sorbed. The impact of this linkage hinges on the amount of water that is in the soil and on the mineralogical composition and particle-size distribution of the solid phase. The physical properties of vadose water may differ at times markedly from those of water that fills the relatively large pores of highly permeable groundwater aquifers because water is a strong dipole and is readily influenced by the net surface charge density of the soil particles and the numbers and kinds of dissolved constituents. Solution ions satisfy the surface charge on soil particles caused by substitution of one element for another in the crystal lattice of clay minerals by several mechanisms. The net surface charge of an assemblage of soil particles gives rise to an electrical field that affects the distribution of cations and anions within the water films. It may also change the configuration and properties of the water close to particle surfaces.

Figure 3 shows distribution of monovalent ions in the soil solution as a function of distance from the soil particle surface within a water-saturated soil pore. For the more concentrated solution of 0.1 N the impact of the electrical field is not evident at distances greater than about 5 nm, while that for the dilute solution (0.001 N) extends further than 20 nm into the pore. The thickness of the electrical "double layer" that neutralizes the excess surface charge of soil particles is not only affected by the total electrolyte concentration, but also by the mineralogical composition of soil particles and by the valency and hydration of ions in the soil solution. The distributions in Figure 3 are for a water-saturated soil. As the water content decreases, the cations and anions are forced to occupy a space limited by the thickness of the water films on the soil particle surfaces. Such a surface-related phenomenon may give rise to swelling pressures, streaming potentials, and salt sieving.

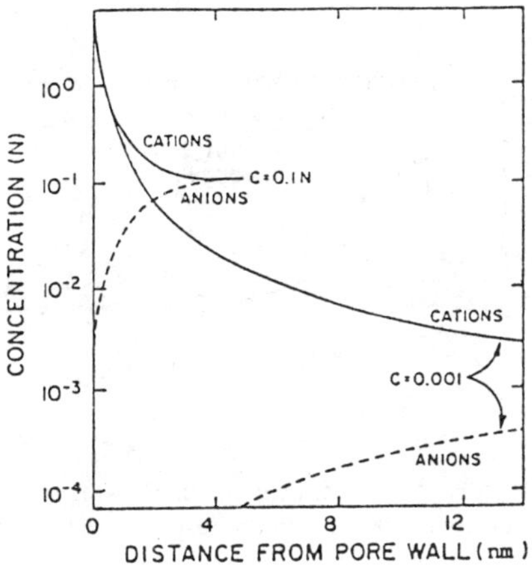

Figure 3. Distribution of monovalent cations and anions near the surface of a clay particle.

Let's illustrate the interplay of the several physical and chemical processes alluded to above with a data set from a laboratory study. A solution containing two tracers (^{36}Cl and $^{3}H_2O$) was applied and leached through water-saturated soil columns under controlled conditions. The relative appearance and position of those tracers in the effluent of the columns differ markedly depending upon the particular condition within the soil. The concentration, pH, and pore water velocity were each independently controlled. In Figure 4, as the concentration of the soil solution decreases, the breakthrough curves of ^{36}Cl shift to the right with their maxima decreasing. The data reveal that at pH 4, chloride is adsorbed. Owing to the fact that an equal number of negative and positive exchange sites exists at pH 3.6, we would expect $^{36}Cl^-$ to be exchanged for its nonradioactive isotopes on the clay surfaces. Differences in shapes and positions of the curves in Figure 4 are therefore a result of the concentration of the soil solution rather than caused by hydrodynamic and geometric aspects of the flow regime.

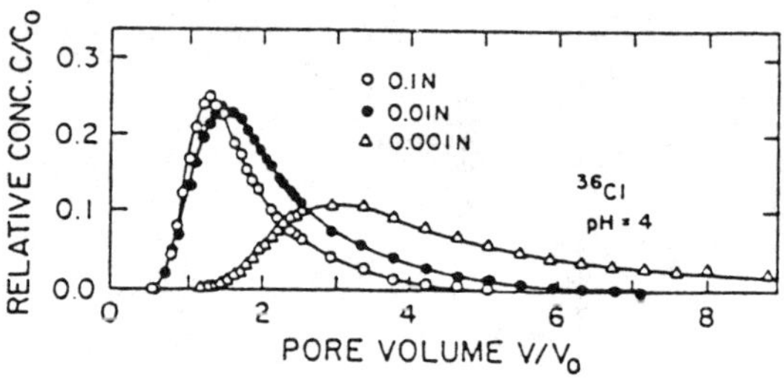

Figure 4. Observed ^{36}Cl effluent curves showing the effect of total ionic concentration on the displacement process.

As the pH of the variability charged soil monotonically increases above 3.6, the relative proportion of negative to positive exchange sites increases. Thus as is shown in Figure 5 for a constant soil solution concentration of 0.001N, the ^{36}Cl breakthrough curve shifts to the left as the pH increases. At pH 9, the early breakthrough of ^{36}Cl is, indeed, indicative of tracer that is repelled from the predominantly negatively charged clay surfaces. When quantifying adsorption-exchange processes, and in view of the experiments shown in Figure 5, it should be remembered that the surface charge characteristics of soil colloids are of two general types, one having a constant surface charge and a variable surface potential and the other having a constant surface potential and a variable surface charge. The charge of the former is permanent and independent of solution concentration. The charge of the latter is determined by the nature of the adsorbed ions, the concentration of the ions in solution, and the pH of the soil. Although arid soils usually are dominated by constant charge colloids and tropical soils by those of constant potential, all soils are mixtures of both, and hence their ion-exchange behavior under conditions that induce major shifts in pH cannot be ignored.

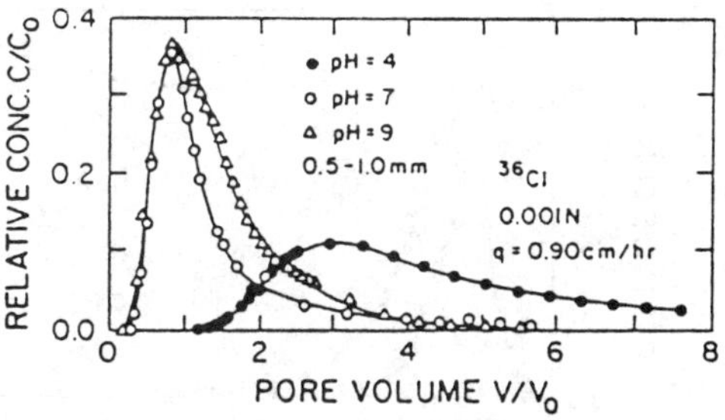

Figure 5. Observed ^{36}Cl effluent curves showing the effect of solution pH on the displacement process.

Figures 4 and 5 pertain to soil columns composed of 0.5- to 1.0-mm aggregates through which water was flowing at a constant flux of 0.9 cm/h. Figures 6 and 7 manifest the impact of water velocity and pore geometry, respectively. The higher flow velocity leads to an earlier breakthrough of the solute in the effluent. Larger aggregates similarly result in earlier breakthrough. Hence both flow velocity and aggregate size affect the transport rate, thus perhaps suggesting that the interactions between the soil and its solution are not at equilibrium under these conditions.

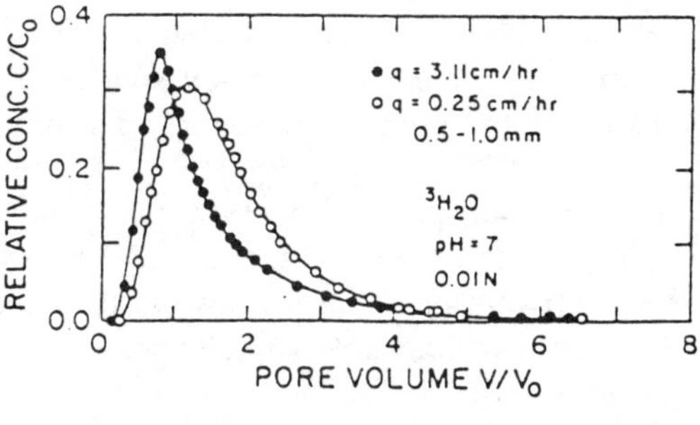

Figure 6. Observed ^{36}Cl effluent curves showing the effect of pore water velocity on the displacement process.

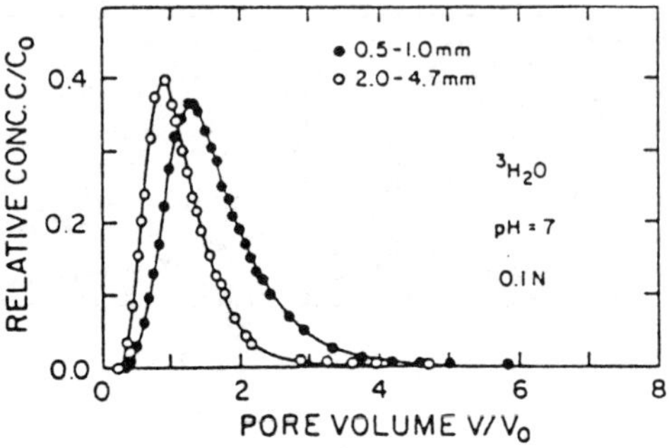

Figure 7. Observed ^3H_2O effluent curves showing the effect of aggregate size on the displacement process.

The data in Figures 4-7 were for a water-saturated soil. Similar data for unsaturated soils are not readily available. Even for water-saturated soils, the displacement of a solute through a soil having different proportions of cationic species (Na^+, Ca^{2+}, Mg^{2+}, etc.) on its exchange sites has not been extensively observed. Although such detailed experimental information is lacking for most soils, there is sufficient evidence in hand to suggest that the separately held conceptual views of the hydrodynamicist and the geochemist should be amalgamated into a more unified theory. It is, indeed, naive to consider that either water or its solutes can be within a field soil without being affected by soil particle surface interactions, or that the behavior of soil solutes can be described without considering the pore water velocity distribution.

Chemical and physical factors that contribute most to changes in the hydraulic conductivity parallel those that alter the extent of the electrical double layer: low electrolyte concentrations, high values of the exchangeable sodium percentage (ESP) or the sodium adsorption ratio (SAR), and increased amounts of expansive 2:1 phyllosilicate minerals (montmorillonite and illite). The effect of total salinity on the unsaturated hydraulic conductivity $K(\Theta)$ is illustrated in Figure 8. A solution of 1000 meq/l $CaCl_2$ percolating through a soil sample resulted in the upper curve yielding values of the hydraulic conductivity more than one order of magnitude greater than those when a water containing only 25 meq/l passed through the soil. Hence, the quality of the water as it passes through the vadose zone, not yet incorporated into any theory of water transport, must be included in our description of vadose water.

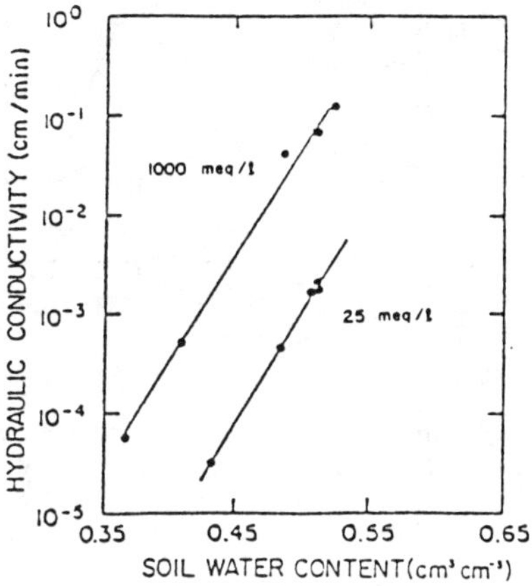

Figure 8. The impact of water quality on the value of the unsaturated hydraulic conductivity.

CHALLENGING ISSUES IN THE VADOSE ZONE

If we ignore, for the moment, osmotic and electrochemical effects as has been nearly the case from Buckingham's time to the present, three challenges persist even after a century of soil physics research. The first is the hysteretic nature of the soil water content versus its energy status. Remember that Buckingham said "the attraction of soil to water--depends in some as yet unknown way, differing from soil to soil, on the water content of the soil". Said today for a given soil, the soil water content is not a unique function of ψ, but depends upon the previous history of the soil. The hysteretic nature of the soil water retention function $\Theta(\psi)$ is illustrated in Figure 9. Hysteresis has important effects on water and solute distributions during field conditions that involve alternate wetting and drying. While its concepts have been described in various textbooks, various attempts to quantitatively describe hysteresis have been less than successful. Despite its importance, it remains an enigma to soil physicists and it has only sparingly been included in the field scale flow and transport models.

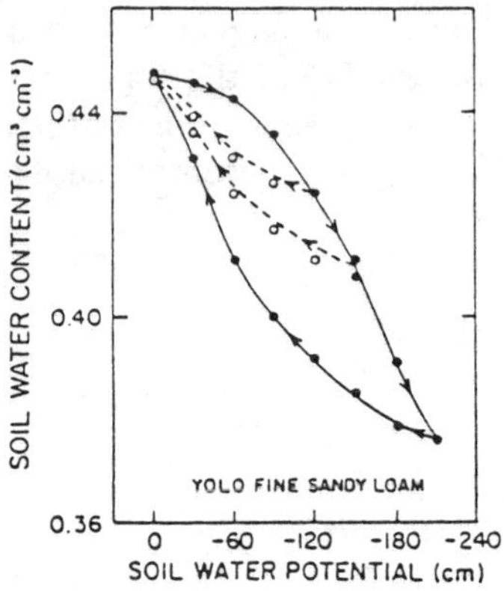

Figure 9. Soil water retention curves. Arrows indicate the direction at which the values of ψ are decreasing or increasing.

The second persistent challenge is to predict the temperature dependency of soil water properties. Compared with hysteresis, the effects of temperature and the hydraulic properties have been largely ignored. Considering that substantial daily and seasonal temperature fluctuations prevail near the soil surface and that the geothermal gradient exists within most of the vadose zone, it is surprising that only a handful of researchers has studied the temperature effect in detail. Figure 10 shows for two soils that at a given value of ψ, less water will be retained when the temperature is high. The influence is especially significant in fine-textured soils. For the Oxford clay, an 8° C temperature increase decreases the soil water content about 5% by weight, which in turn, corresponds to a 10-fold decrease in the hydraulic conductivity. The hydraulic conductivity, already known to be strongly dependent upon soil water content, is very sensitive to soil temperature (Figure 11). Several researchers have sought to explain the temperature dependence on these soil water properties through the temperature dependency of the viscosity and surface tension of water. Conflicting evidence exists and should motivate additional work on this important but neglected area of research.

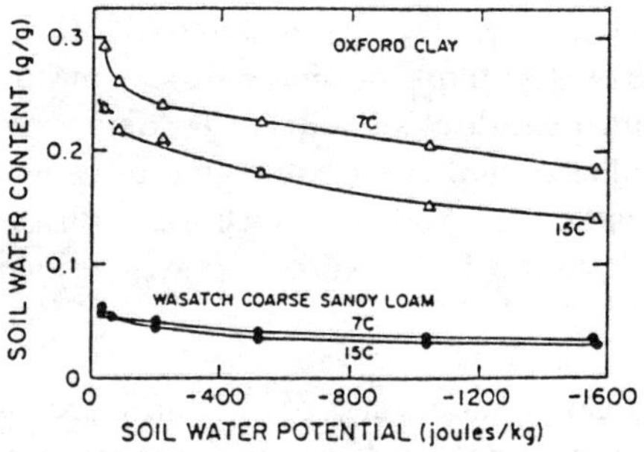

Figure 10. Soil water retention curves for two soils at two temperatures [after Taylor, 1958].

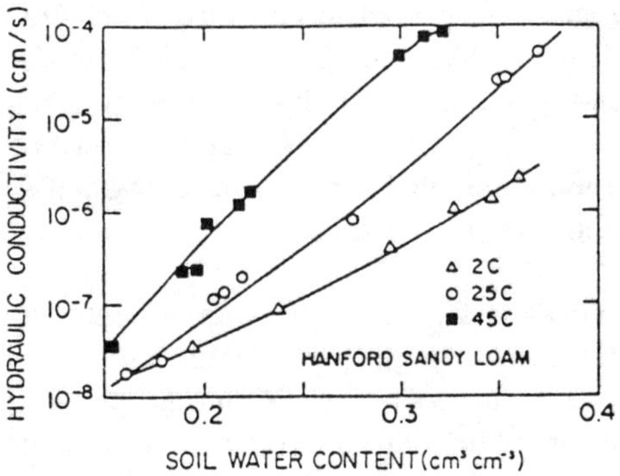

Figure 11. Unsaturated hydraulic conductivity $\underline{K}$ as a function of water content Θ at three temperatures [after Constantz, 1982].

The third challenge, closely associated with soil temperature, is the transport of water in both vapor and liquid phases under isothermal and nonisothermal conditions. In unsaturated soil, thermal gradients are efficient movers of water in the vapor phase. As water vaporizes into the gaseous phase, the concentration gradients in the water films of the vadose zone are capable of moving significant quantities of water. To date only a handful of experimental studies treat the simultaneous effects of thermal, solute and hydraulic gradients. And with such a dearth of experimental observations, theoretical concepts remain incomplete and underdeveloped.

The transport and fate of contaminants and naturally occurring elements in the vadose zone are also related to the metabolism and energetics of microbial communities in that subsurface environment. The importance of subsurface microorganisms in controlling or mediating the quality of groundwater has only recently become recognized as important. Soil microbiologists, particularly those in the agricultural sciences, accustomed to nutrient-rich environments of the topsoil, believed that the presence and activity of microorganisms in the vadose zone was either extremely limited or nonexistent. Recent studies within this decade have shown that organic contaminants can be transformed by subsurface microorganisms. Such transformations offer several scientific challenges to the hydrologic community. A better understanding of the growth and metabolism of microbial colonies within the vadose zone as water and its constituents migrate slowly through their domains from the sun-energized soil surface to the water table has yet to be explored.

With today's computer technology ranging from improved personal computers to huge mainframes, solutions of partial differential equations analogous to Richards' equation and those similarly derived for solute and contaminant transport in unsaturated soils are relatively abundant. Do not mistake their abundance with correctness or their successful application to field situations. Many of the solutions are applicable to highly simplified systems that do not embrace real conditions encountered in the field, while others, that conceptually include the more important features of the vadose zone, are still waiting to be fitted with numerical values of parameters that remain ambiguous or unavailable owing to a lack of field technology to estimate them. Indeed, there exists no unified field technology to measure the "capillary conductivity" parameter of Buckingham. We have yet to agree on preferred methods of analyzing soil water properties even in the laboratory.

The development of new methods, both experimental and theoretical, has been slowed during the past two decades by the recognition of the inherent spatial variability of field soils and the vadose zone. It was in the 1960's, or that period which I call the beginning of environmental citizenship in the U.S., that we started to use equations derived at the laboratory scale to predict the movement of water and its dissolved constituents in waste disposal sites, farmer's fields or larger regions. It was at that time that we attempted to theoretically describe transport in the vadose zone owing to our need to understand the management of the quality of water moving from so-called non-point discharges. Laboratory soil columns traditionally used to ascertain soil water properties could not substitute for field scale validation. Two things happened. Money for field scale hydrologic research dried up during the past 20 years, and calculators improved, leaving us today with many field scale simulations and a dearth of observations to validate our new ideas.

Our potential for theoretical progress is exciting. Various stochastic and statistical approaches are being developed and the more traditional deterministic ideas continue to evolve. Stochastic and statistical models describe the variables as random functions, which depend on the distribution of values of the soil properties which determine water and solute movement. Rather than predict values of water content of solute concentration as a function of position and time, stochastic models predict expected averages and variances, and are used to calculate the probability of having a given value appear at a given depth or time. Many investigators are pursuing these stochastic opportunities in the vadose zone. A number of scientists continue to address deterministic concepts following the early work of Philip (1955, 1957) who provided much of the physical and mathematical groundwork for infiltration into the vadose zone. It is indeed exciting that Professor Arthur W. Warrick and his colleagues here at U. of A. reported this year the first analytic solution of Richards' equation for drainage and redistribution of water in the vadose zone. A first in the last half-century!

Experimental opportunities also exist for the vadose zone. No one has yet measured the water content or its potential energy within a single soil pore. The possibility of measurement becomes greater as tomography and differential tomography using γ and X-ray radiation, nuclear magnetic resonance and other energy sources are applied to soil. These measurements will allow a microscale examination of the physics of water flow in films along soil particle surfaces and fractured rock, and hopefully reveal how the concentration and distribution of chemical constituents in the soil solution alters the macroscopically conductivity and soil water characteristic curve, the effects of temperature on the hydraulic properties, the displacement of air during infiltration or drying, the simultaneous movement of water vapor and transport of heat, the impact of localized macropore geometries on leaching efficiency during infiltration, and the quantification of soil water properties for soils that shrink and swell or crack and consolidate upon drying and wetting. Other approaches in instrumentation focused at the pore or small sample scale are undoubtedly being developed. A vigorous effort must also be made to collect field data at a much larger scale from carefully designed experiments to aid in the continued development of new theories for the vadose zone.

CONCLUDING REMARKS

In conclusion, I would like to emphasize that the relative importance of the vadose zone is much greater in western U.S. where irrigated agriculture is practiced. Irrigated agriculture has produced food for civilizations that have come and gone during the past thousands of years. Today, roughly 25% of the value of all crops in the U.S. comes from irrigation on 10% of the land farmed. Over 24 million hectares of irrigated land produces more than 50% of our vegetables, 56% of our potatoes and the majority of our fresh fruit and nuts. In fact, irrigated regions of western U.S. produce 40% of the table food consumed in the country as well as that exported to other countries. There is no doubt of the benefits of irrigated agriculture that assures abundant yields, short-term security against droughts and diverse opportunities for farm management that guarantees a return on private investment with benefits for the producers, processor, distributor and consumer.

On the other hand, each irrigated acre often uses 1 million gallons of water - water that could be used in other ways for the benefit of society. Moreover, irrigation always degrades water even with proper management. With proper management, including natural or constructed drainage systems, the drainage water traveling through the vadose zone carries unwanted salts and agrochemicals and their metabolites that require "disposal". Where are these "disposal sites" for the drainage of irrigated agriculture? The oceans, inland seas, deep underground storage, evaporation ponds, or their intentional or unintentional dilution with high quality waters have all been categorically rejected by one or more sectors of our society.

It is no longer taken for granted that the use of land and water for irrigated agriculture is the preferred use. Permanent irrigated production of food and fiber demands two sacrifices of some value elsewhere. Firstly, the storage and diversion of water for agriculture reduces the water available for other uses. Diversion of water not only modifies the original ecosystem but affects the recreational and economic value of the water as well as other resources. Secondly, the waste product of irrigated agriculture - degraded drainage water containing unwanted pollutants and salinity - must be conditionally accepted and paid for - by somebody! Irrigated agriculture must adapt to changing physical and social conditions in order to survive. Greater environmental and global values and direct consideration of good quality water for municipalities and industries must be included in the management matrix of irrigated food production.

Perhaps it is now time to reconsider the task given to me thirty years ago. How can I irrigate those vegetables not for maximum yield, but for a sustainable production managing the quantity and quality of water as it passes through the vadose zone into reusable ground water? I call on each of you through our mutual research and education programs to explore the last hydrologic frontier - the vadose zone.

LITERATURE CITED

Briggs, L.J. The mechanisms of soil moisture. Bulletin 10, USDA Bureau of Soils, Washington, D.C. 1987.

Buckingham, Edgar. Studies of movement of soil moisture. USDA Bureau of Soils Bulletin No. 38, Washington, D.C., 1907.

Constantz, J. Temperature dependence of unsaturated hydraulic conductivity of two soils, Soil Sci. Soc. Am. J., 46(3):466-470 1982.

Philip, J.R. Numerical solution of equations of the diffusion type with diffusivity concentration-dependent, Trans. Faraday Soc., 51:855-893, 1955.

Philip, J.R. The theory of infiltration. I. The infiltration equation and its solution, Soil Sci. 83:345-357, 1957.

Richards, L.A. Capillary conduction of liquids in porous mediums, Physics 1:318-333, 1931.

Taylor, S.A. The activity of water in soils, Soil Sci., 86:83-90, 1958.

DONALD R. NIELSEN
Department of Land, Air and Water Resources
University of California
Davis, California 95616

The professional career of Donald R. Nielsen, a native of Arizona, began in 1958, after he obtained the Ph.D. degree in Soil Physics from Iowa State University. Earlier, he had earned the M.Sc. degree in Soil Microbiology as a Paul Steere Burgess Fellow and the B.Sc. degree in Agricultural Chemistry and Soils at the University of Arizona.

From 1958 to the present, Dr. Nielsen has been employed at the University of California, Davis. Having become Professor of Soil and Water Science in 1968, he has served as Director of the Kearney Foundation of Soil Science 1970-75, Director of Food Protection and Toxicology Center 1974, Chair of the Department of Land, Air and Water Resources, 1975-77, Associate Dean of the College of Agricultural and Environmental Sciences 1970-80, Executive Associate Dean 1986-89, and is presently Chair of Agronomy and Range Science. He was a senior post-doctoral fellow, NSF, 1965-66.

Dr. Nielsen's major scientific and professional interests embrace laboratory and field investigations of leaching and miscible displacement of inorganic and organic solutes in the vadose zone, soil water properties, pesticide behavior in soils, soil microbiological transformations during leaching, and geostatistical concepts in relation to the technology of sampling field soils. He has coauthored more than 250 publications and has lectured in more than 20 countries. He has served on editorial boards of eight scientific journals and has been Editor of Water Resources Research (1985-89).

In recognition of his achievements, Dr. Nielsen is a Fellow of the American Geophysical Union, American Society of Agronomy and Soil Science Society of America. He is past president of the Soil Science Society of America, and currently serves as President of the American Society of Agronomy (1990-91) and President of the Hydrology Section of the American Geophysical Union (1990-92). In 1986 he was awarded an Honorary Doctorate of Science by Ghent State University, Belgium.

FOREWORD

Hydrological phenomena, those mysterious transformations of precipitation into runoff, have perplexed humanity for a long time. The interplay of atmospheric dynamics and the passive lithosphere, the incessant throbbing of living organisms — plants and animals — and mainly plants, on this planet which utilize water and return it to nature, result in a sculptured landscape through which flow silvery ribbons. Each ribbon drains a topographical basin, joins adjacent ribbons which together flow into streams. Ultimately, most rivers flow into the ocean.

The chain of events evaporation, precipitation, evapotranspiration, runoff, erosion, deposition, percolation, infiltration, streamflow — involve, as all natural phenomena do, changes and movement of matter and use of energy. The radiant energy streaming ceaselessly from the sun is responsible for the climate on earth, which is the basis of all hydrological phenomena. Thus, we may view hydrological phenomena that we observe today as the result of an interaction between climate as the forcing function, geology as a basic constraint, and man (generally speaking) as major modifying agent. The study of man's activities as a modifying agent is the realm of Water Resources, and several Kisiel Memorial Lectures have dealt with some of its aspects, e.g., the Third (Water Management in the United States), the Sixth (Groundwater Contamination), and the Seventh (Efficiency Gains from Building Equity into Water Development). The Tenth Kisiel Memorial Lecture discusses some fundamental issues arising from the interaction between the forcing function and the basic constraint.

The outcome of the interaction between climate (the forcing function) and geology (the basic constraint) manifests itself in the various landforms, within a broad range of magnitudes, that create the landscape. The landform of interest in hydrology and water resources is the watershed (or catchment) which, in essence, is a transportation system that conveys through a drainage network the runoff and the sediment produced in a basin. The forcing function — the climate — is an expression of the energy which, to a large extent, is responsible for the three-dimensional shape and structure of the drainage network. Nature, so it seems, is parsimonious: landscapes and river basins develop and evolve with a minimum expenditure of energy.

Nathan Buras

REFLECTIONS ON THE 3-DIMENSIONAL STRUCTURE OF RIVER BASINS: ITS LINKAGE WITH RUNOFF PRODUCTION AND MINIMUM ENERGY DISSIPATION

Introduction

It is a great pleasure and honor to deliver the Tenth Kisiel Memorial Lecture and I thank the Department of Hydrology and Water Resources for their kind invitation to be here today. This event is quite special for me because my family and I were good friends of Chester Kisiel, a man whom I always respected for his academic standards and personal honesty. I hope he would have enjoyed the theme of my lecture. In fact I remember many conversations with Chester in which he would always press for a healthy mixture of physical insight and mathematical analysis. This is the spirit in which today's presentation is made; no doubt he would have raised many questions and quite likely would have disagreed on many issues. Nevertheless, in scientific research, it is better to be wrong but interesting, than right and dull . . .I hope Chester would have found it interesting.

Socialism and Capitalism in River Basins

Well developed river basins are made up of two interrelated systems: the channel network and the hillslopes. The hillslopes control production of runoff which in turn is transported through the channel network towards the basin outlet. Every branch of the network is linked to a downstream branch for the transportation of water and sediment but it is also linked for its viability — through the hillslope system — to every other branch in the basin. Hillslopes are the runoff producing elements with the network connectors transforming the spatially distributed potential energy arising from rainfall in the hillslopes to kinetic energy in the flow through the channel reaches. It is precisely the need for effective connectivity that leads to the tree-like structure of the drainage network. Figure 1, from Stevens (1974), illustrates this point. Assume one wishes to connect a set of points in a plane to a common outlet and, for illustration purposes, assume that every point is equally distant from its nearest neighbors. Two extreme ways to establish the connection would be through the spiral and explosion types of patterns. The explosion pattern has the advantage in that it connects every parcel of the system to the outlet in the most direct manner. Nevertheless it rejects any kind of interaction between the different parts and the total path length for the system as a whole is extremely large. Thus, although it has the minimum average path connecting each parcel to the outlet, it lacks shortness as a whole. The spiral pattern, on

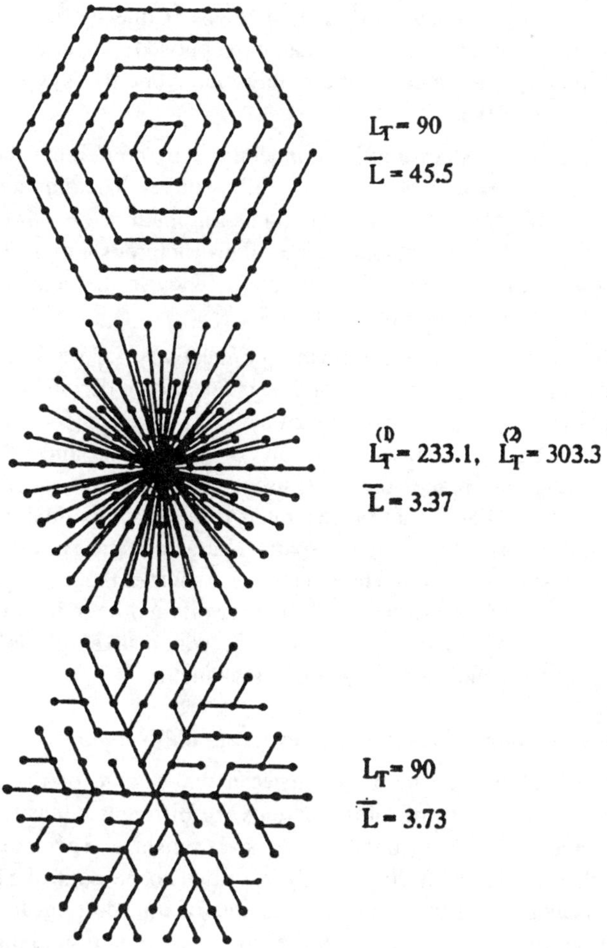

Figure 1. Different patterns of connectivity of a set of equally spaced points to a common outlet. LT is the total length of the paths, and L is the average length of the path from a point to the outlet. In the explosion case LT(2) refers to the case when there is a minimum displacement among the points so that there is a different path between each point and the outlet [from Stevens, 1947].

the other hand, is quite short for the system as a whole, but it leads to an extremely large average path from a point to the outlet. One is tempted to say that, from an organization point of view, the spiral represents pure socialism and the explosion pure capitalism. In one case the system is supposed to operate at its best as a whole with a total disregard the average individuals who find themselves in the worse possible condition. In the other case, individual are supposed to operate at their best, completely oblivious to their neighbors, but the system as a whole cannot survive.

Branching patterns accomplish connectivity by combining the best of the two extremes, they are short as well as direct. The drainage network, as well as many other natural connecting patterns, is basically a transportation system for which the tree-like structure is a most appealing structure from the point of view of efficiency in the construction, operation and maintenance of the system.

The drainage network accomplishes connectivity for transportation in three dimensions working against a resistance force derived from the friction of the flow with the bottom and banks of the channels, the resistance force being itself a function of the flow and channel characteristics. This makes the analysis of the optimal connectivity a complex problem that cannot be separated from the individual optimal channel configuration and from the spatial characterization of the runoff production inside the basin. The question is whether there are general principles that relate the structure of the network and its individual elements with the rate of energy expenditure which takes place in the system as a whole and in each one of its elements.

Principles of Energy Expenditure in Drainage Networks

A link of a drainage network carries a wide range of discharges resulting from a variety of rainfall events (of different intensities and duration) and antecedent conditions of soil moisture. The individual channel characteristics are commonly assumed to be controlled by a bankful discharge that the channel is capable of transporting. It is also true, though, that most of the work the flow performs throughout time takes place at discharges smaller than the bankful capacity. From this point of view the mean annual flow may be considered a more representative discharge condition to characterize the work being done by the flow. Thus it is likely that any principles of optimal energy expenditure responsible for the three-dimensional structure of the drainage network will yield the same type of results when applied to the case of bankful discharges as when the flow is characterized at every link by the corresponding mean annual value.

Three different principles are now postulated:

1. The principle of minimum energy expenditure in any link of the network.

2. The principle of equal energy expenditure per unit area of channel anywhere in the network.
3. The principle of minimum energy expenditure in the network as a whole.

It will be shown that the coupling of these principles is sufficient explanation for the tree-like structure of the drainage network and, moreover, that they explain many of the most important empirical relationships observed in the internal organization of the network and its linkage with the flow characteristics. The first principle expresses a local optimal condition for any link of the network. The second principle expresses an optimal condition throughout the network regardless of its topological structure which later on in this paper will be in a probabilistic framework. It postulates that energy expenditure is the same everywhere in the network when normalized by the area of the channel on which it takes place. Thus, even with the first principle, there will be channels that spend much more energy per unit time than others only because of their larger discharge. The second principle makes all channels equally efficient when adjusted for size. The third principle is addressed to the topological structure of the network and refers to the optimal arrangement of its elements.

The first principle is similar to the principle of minimum work used in the derivation of Murray's law in physiological vascular systems. C.D. Murray in 1926 derived a relation which states that the cube of the radius of a parent vessel should equal the sum of the cubes of the radii of the daughter vessel (e.g., see also Sherman, 1981). He assumed that two energy terms contribute to the cost of maintaining blood flow in any vessel: (a) the energy required to overcome friction as described by Poiseuille's law, and (b) the energy metabolically involved in the maintenance of the blood volume and vessel tissue.

Minimization of the cost function leads to the radius of the vessel being proportional to the 1/3 power of the flow. Uylings (1977) has shown that when turbulent flow is assumed in the vessel, rather than laminar conditions, the same approach leads to the radius being proportional to the 3/7 power of the flow. The second principle was conceptually suggested by Leopold and Langbein (1962) in their studies of landscape evolution. It is of interest to add that minimum rate or work principles have been applied in several contexts in geomorphic research. Optimal junction angles have been studied in this context by Howard (1971), Roy (1983), and Woldenberg and Horsfield (1986) among others. Also the concept of minimum work as a criterion for the development of stream networks has been discussed under different perspectives by Yang (1971), and Howard (1990) among others.

Energy Expenditure and Optimal Network Configuration

Consider a channel of width w, length L, slope S, and flow depth, d. The force responsible for the flow is the downslope component of the weight, $F_1 = \rho$ gdLw sinβ = ρ gdLwS where sinβ ≃ tanδ = S. The force resisting the movement is the stress per unit area times the wetted perimeter area, $F_2 = \tau$ (2d + w) L, where a rectangular section has been assumed in the channel. Under conditions of no acceleration of the flow, $F_1 = F_2$, and then τ = ρ gSR where R is the hydraulic radius $R = A_w/P_w$ = wd(2d + w),Qw Λ Pw being the cross-sectional flow area, and the wetted perimeter of the section respectively. In turbulent incompressible flow the boundary shear stress varies proportionally to the square of the average velocity, $\tau = C_f \rho V^2$, where C_f is a dimensionless resistance coefficient. Equating the two expressions for τ, one obtains the well known relationship, $S = C_f v^2/(Rg)$, which gives the losses due to friction per unit weight of flow per unit length of channel. There is also an expenditure of energy related to the maintenance of the channel which may be represented by F(soil,flow) • P_w • L where F(•) is a complected function of soil and flow properties representing the work per unit time and unit area of channel involved in the removal and transportation of the sediment which otherwise would accumulate in the channel surface. From the equations of bed load transport one may assume that $F = K \tau^m$ where K depends on the soil and fluid properties and m is a constant.

In channel of length L and flow Q the rate of energy expenditure may then be written as

$$P = C_f \rho \frac{v^2}{R} QL + K\tau^m P_w L = C_f \rho P_w \frac{Q^3}{A_w^3} L + KC_f^m \rho^m v^{2m} P_w L \quad (1)$$

The coefficient C_f depends mainly on the channel roughness which tends to decrease only slightly in the downstream direction; on the whole the downstream reduction in roughness resulting from a decrease in particle size is compensated by other forms of flow resistance like that offered by bars and channel bends (Leopold et al., 1964). According to the second principle of energy expenditure, $P_1 = P/(P_w \bullet L)$ is the same anywhere in the network. Substituting P from equation (1), one obtains

$$P_1 = C_f \rho\, v^3 + KC_f^m \rho^m v^{2m} = \text{constant} \quad (2)$$

which implies that the velocity tends to be constant throughout the network. This has been corroborated by the field investigations of Leopold and Maddock (1953), Wolman (1955), and Brush (1961) who obtained values of $z < 0.1$ in the downstream relation between velocity and discharge, $v = CQ^z$, this being the case for both mean annual flow conditions or bankful discharges throughout the network. Also the field experiments of Pilgrim (1977) corroborate this finding, although as pointed out by Howard (1990) this may not be a universal kind of behavior. Substituting the width $w = Q/(vd)$ in Equation (1) one gets

$$P = \frac{QL}{d}\left[C_f \rho v^2 + KC_f^m \rho^m v^{2m-1}\right] + dL\left[2C_f \rho v^3 + 2KC_f^m \rho^m v^{2m}\right] \quad (3)$$

the terms in brackets being constant throughout the network for a given flow condition. According to the first principle of energy expenditure, P should be a minimum in any link of the network. If the link is transporting a discharge Q, this means $dP/d(d) = 0$ which yields

$$Q = \text{constant} \bullet d^2 \text{ or } d = \text{constant} \bullet Q^{0.5} \quad (4)$$

Thus in any link of a network the mean annual flow or the bankful discharge are proportional to the square of their corresponding flow depths, the constant of proportionality being the same everywhere in the network. The above result has been observed by field investigators. Leopold, et al. (1964) found $d \sim Q^f$ with $f \simeq 0.4$ for the dependance of depth on flow in the downstream direction, with the same exponent valid both for bankful conditions and for mean annual flow conditions. Using Equation (4) in the expression for width, $w = Q/(v \bullet d)$,gives

$$w = \text{constant} \bullet Q^{0.5} \quad (5)$$

which says that in the downstream direction the width varies proportionally to the square root of the discharge. Leopold, et al. (1964) found a very good relationship between width and the square root of the discharge in the downstream direction for both bankful and mean annual flow conditions.

Substituting Equation (4) in Equation (3) we obtain the optimal power expenditure at any link as

$$P = kQ^{0.5}L \quad (6)$$

Adding over all links of the network we obtain the total rate of energy expenditure under optimal conditions

$$E\sum_i P_i = k\sum_i Q_i^{0.5}L_i \tag{7}$$

where **k** varies with the discharge but is constant throughout the network if mean annual flow or bankful conditions are operating throughout the basin. In an explosion pattern like the one of Figure 1, the Q'_is are small since there is no aggregation of flows from tributary links; on the other hand, the sum of L'_is is extremely large and so is E. If each node in Figure 1 has a constant discharge, Q, then

$$E=kQ^{0.5}L_T \tag{8}$$

where L_T is the total path length. In the case of the spiral pattern with a constant discharge at every node, one has

$$E = kLQ^{0.5}[1 + 2^{0.5} + 3^{0.5} + \ldots + N^{0.5}] \tag{9}$$

where L is the constant distance between neighboring points. Although L_T is small for the spiral, E is again prohibitively large. On the other hand the tree-like pattern combines a piecewise aggregation of flows throughout the system at the same time that it keeps quite short the total length of the flow paths. This yields a much smaller total rate of energy expenditure, E. In the case of Figure 1, if the input flow at any node is taken as equal to one, the corresponding values of E are spiral = 574k; explosion = 303k; and tree-like = 151k. The explosion pattern is only relatively competitive when most of the points are close to the common outlet. If one keeps adding points further away from the outlet, the total length of the explosion pattern increases dramatically and so does the total energy expenditure E. The above comparison although illustrative only correct if one assumes that k is the same in all cases which implies that the flow velocity has remained the same in all cases which is not necessarily true in natural networks.

Horton's Laws: The Role of Chance and Necessity

The fantastic variety of forms and shapes of drainage networks embodies a deep sense of regularity in formal relations among the parts, an important example of which are the empirical laws founded by R.E. Horton in 1945. Horton's law of stream numbers involves the relative arrangement of streams which he stated as: "the number of streams of different orders in any given drainage basin tends closely to approximate an inverse geometric series in which the first term is unity and the ratio is the bifurcation ratio" (Horton, 1945). Mathematically it is expressed as

$$\frac{N_w}{N_{w+1}} = R_b \tag{10}$$

where N_w is the number of streams of order w, and R_b is the bifurcation ratio, N_w is the usually estimated with the Strahler ordering procedure: (l) channels that originate at a source are defined as first order streams; (2) when two streams of order w join, a stream or order w + 1 is created; and (3) when two streams of different order join, the channel segment immediately downstream, has the higher order of the two combining streams. $N_\Omega = 1$ where Ω is the order of the basin network which for a fixed number of sources, N_1,is a measure of the degree of branching. Horton's law of stream lengths is expressed as

$$\frac{L_{w+1}}{L_w} = R_L \tag{11}$$

where L_w is the average length of streams of order w and R_L is the length ratio.

In 1966 R.L. Shreve provided a statistical interpretation of Horton's law of stream numbers. He defined a topologically random population of channel networks as a population within which all topological distinct networks with a given number of links, or equivalently with a given number of sources, are equally likely. Topologically distinct networks are those whose schematic map projections cannot be continuously deformed and rotated in the plane of projection so as to become congruent. Shreve (1966) proposed that in absence of geological controls a natural population of channel networks will be topologically random. He noticed that inherent in the definition of stream order is the corollary that no arborescent network can depart very far from Horton's geometric series law. The fact that for every stream of given order, except the first, there must be at least two streams of the next lower order means that on a Horton diagram of log N_w – vs – w, the points for any channel network with given order and given number of sources "will necessarily lie within a relatively restricted parallelogram-shaped region whose long diagonal is the locus of points which exactly satisfy Horton's law". Moreover it is around the diagonal of the narrow parallelogram where most of the networks with given order and number of sources will be located. Thus the fact that natural networks tend to fulfill Horton's law of stream numbers is really a consequence of the ordering system. Most intriguing is the fact that drainage networks tend to have a bifurcation ratio, R_b, close to 4 with most values lying in a small range between 3 and 5. Shreve showed that in a topologically random population of networks with a given number of sources, the most probable network order Ω is that which makes R_b closest to 4.

In 1969 Shreve provided a statistical interpretation of Horton's law of stream lengths. Taking the link lengths as random variables with a common distribution he showed that networks tend to follow Horton's law of stream lengths, the most probable networks giving the straightest lines in the plots of log L_w – vs – w with an R_L of approximately 2. This agrees well with the values observed in nature which lie between 1.5 and 3.5. LaBarbera and Rosso (1987) and Tarboton, et al. (1988) have shown that R_b and R_L are connected through the fact that drainage networks exhibit a fractal structure with fractal dimension given by

$$D = \frac{\log R_b}{\log R_L} \tag{12}$$

The measurements of Tarboton, et al. (1988) indicate that these networks are space filling with D = 2 which in turn implies $R_b = R_L^2$.

After Shreve's (1966, 1969) classical papers, it has been commonly assumed by hydrologists and geomorphologists that the topological arrangement and relative sizes of the streams of a drainage network are just the result of a most probable configuration in a random environment. Thus the value $R_b = 4$ and the implied $R_L = 2$ are explained solely as being those with the highest probability of occurrence. We believe that in an evolutionary system like the drainage network both chance and necessity should be operating; moreover that the influence of necessity is felt through a tendency to minimize the total rate of energy expenditure in the network and the rate of energy expenditure per unit area of channel anywhere in the network. According to the third principle of optimal energy expenditure, the topological arrangement of the network elements in this space will be such as to lead to the minimization of the total rate of energy expenditure, E. It will be now shown that the third principle leads to drainage networks which obey Horton's laws of stream numbers and stream lengths with the values of R_b and R_L typically observed in nature.

A square grid is superimposed on a geometrical region, the problem being to find the network which minimizes the function E. The optimization implemented is similar to the strategy developed by Lin (1965) for the traveling salesman problem. It starts with an initial configuration which is perturbed by randomly changing the flow direction of a randomly chosen node among its eight surrounding neighbors under the constraint that the network should drain the whole area to a common outlet, no lakes being allowed inside the basin.

The change on E is computed between the new and the old configuration, and if it is negative, the new configuration is adopted as base configuration and the process is iterated. On the other hand, if the change on E is positive, the old configuration is perturbed again. The procedure leads to a network in which no improvement on E appears after many perturbations. The entire process is repeated many times to obtain a set of networks with different minimal E. The configuration with the overall minimum $\Sigma_i P_i$ is selected as the solution to the problem. Although there is no assurance that the solution represents a global optimum, experience shows that the procedure is quite effective in obtaining solutions near the global optimum starting from arbitrary initial configuration. In the computation of the total rate of energy expenditure the discharge Q through any link is taken as proportional to the total area draining through the link. Figure 2 shows examples of the initial configurations that were chosen. The final results constitute in what will be called optimal channel networks (OCN's). They satisfy in all cases Horton's laws with $R_b \approx 4$ and $R_B \approx 2$. OCN's also satisfy a variety of well known empirical relationships like Melton's and Moon's laws, Hack's relation as well as the observed distribution of link lengths which is a power law with exponent close to 2.0 (Tarboton, et. al, 1988). Moreover OCN's exhibit a power law probability distribution for discharge throughout the population of links which have in all cases an exponent of -0.45 which is the same as the one observed in natural basins. This topic will be addressed later in this lecture. Figure 3 presents the OCN resulting from one of the initial conditions shown in Figure 2 as well as the comparative results of the structures of the initial and final configuration.

Scaling Implications of Optimal Energy Expenditure

The rate of energy expenditure per unit area of channel, P_1, may be written as

$$P_1 = \frac{\rho g Q S}{(w + 2d)} + K C_f^m \rho^m v^{2m} \tag{13}$$

Substituting for w and d the expressions obtained from the joint application for the two principles of energy expenditure, $d \propto Q^{0.5}$ and $w \propto Q^{0.5}$, one obtains (with constant velocity throughout the network),

$$P_1 + c_1 Q^{0.5} s + c_2 \tag{14}$$

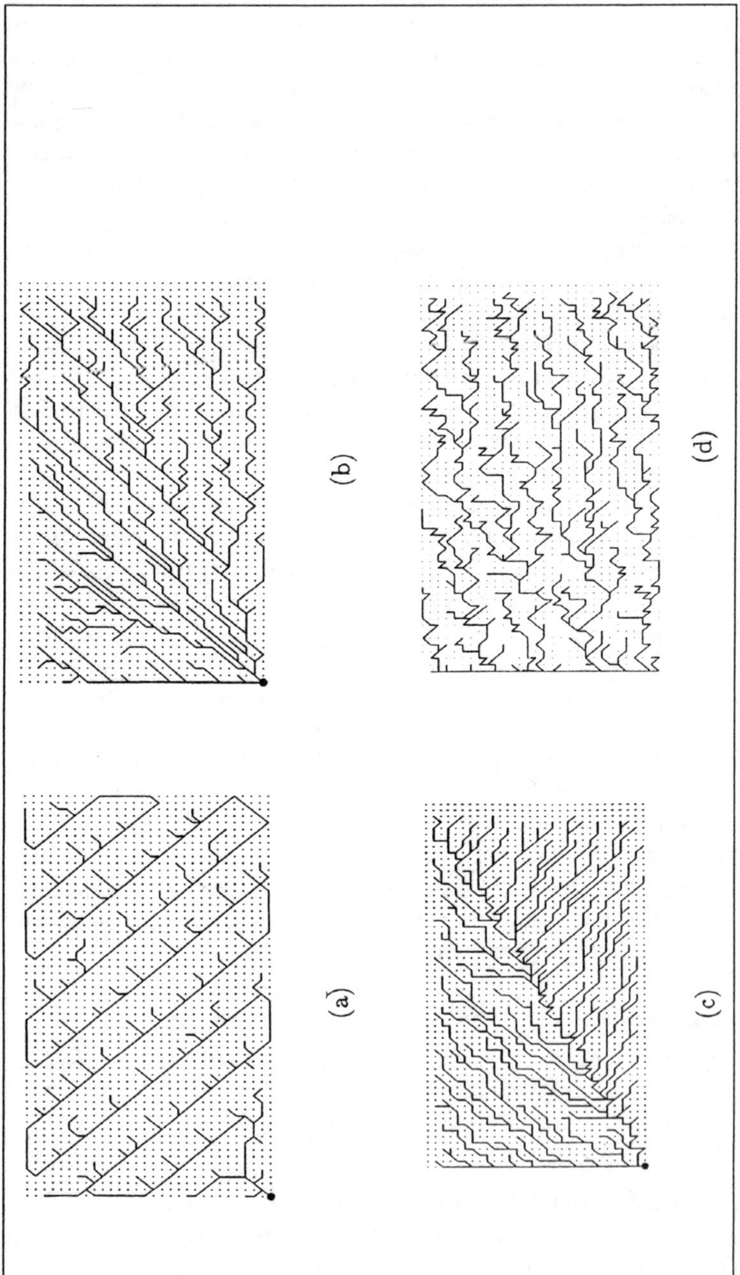

Figure 2. Four types of initial conditions for the 30 x 60 grid optimization.

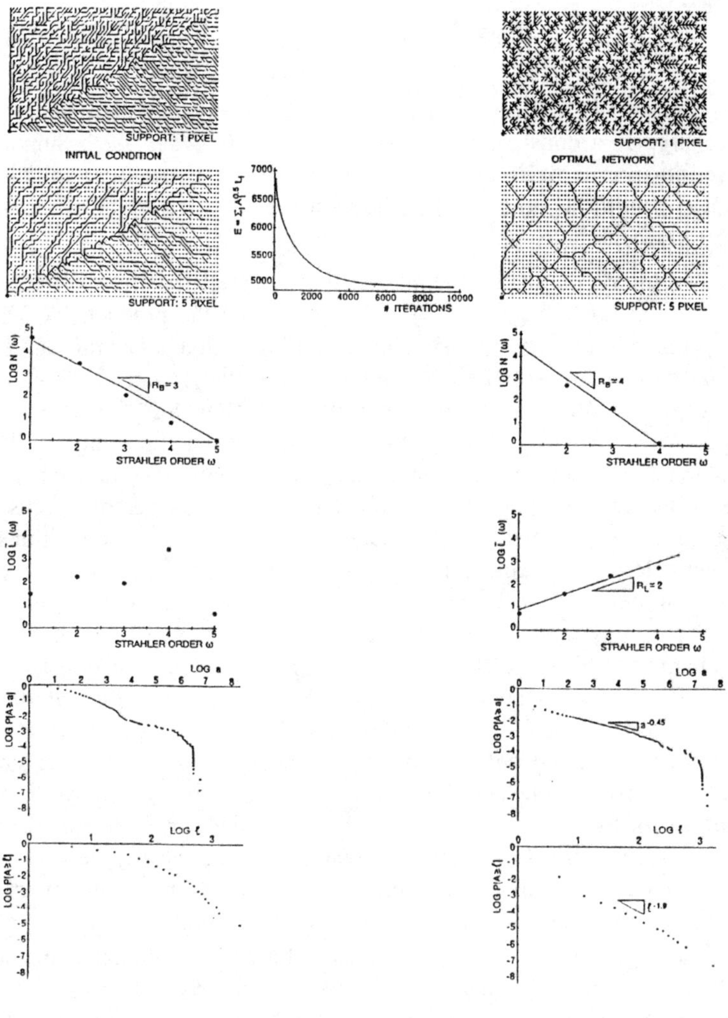

Figure 3. Structural characteristics of the initial network compared with those of the optimal network obtained from minimizing energy expenditure. Rate of energy expenditure in the initial condition is 6843 units, after 55,000 iterations it has stabilized in 4,942 units. Streams are identified as those pixels with cumulative contributing area greater than a support threshold, networks are shown here for support thresholds of 1 and 5 pixels, results above are for 1 pixel, case which essentially are the same as for the 5 pixels support.

For a given flow condition c_1 and c_2 are constant throughout the network and under the second principle P_1 is constant in all links. Thus

$$Q^{0.5}S = \text{constant} \tag{15}$$

implying that the product of the slope times the square root of the discharge is a constant throughout the network when mean annual flow conditions exist everywhere in the basin. Under bankful discharge conditions, this will still be valid with a different constant on the right-hand side.

Studies of the stable channel section of gravel rivers by Parker (1978) and Ikeda et al. (1988) have shown that the product $d \bullet S$ is only a function of the sediment characteristics, with $d \propto Q^{0.5}$ this agrees with the previous result. Leopold and Wolman (1957) report a large number of rivers in the United States and India that show $SQ^{0.44}$ = constant. Leopold et al. (1964) also reports the values for the exponent z in the relation $S \propto Q^z$ for observations in the downstream direction. An average value of $z = -0.49$ was obtained for streams in the midwestern United States for both bankful and mean annual flow conditions. Nevertheless for ephemeral streams in semi-arid regions, they quote an exponent closer to –1.0. The above field data are probably quite unreliable for studying the relationship SQ^z. One needs to measure both the discharge and the slope along individual links. The identification of the network itself is not a trivial matter, and it is only recently through digital elevation maps (DEM's) that the network with the slopes of its links and their individual contributing areas has been objectively studied. DEM's consist of elevations in a rectangular grid with, usually, 30 m spacing. In the U.S.A. grids with 30 m to a side are common. Each grid block is termed a pixel and streams are then usually defined as those pixels with cumulative drainage area greater than a support area threshold (O'Callaghan and Marks, 1984; Band, 1986).

Discharge measurements in every link of a network are not available and since the mean annual flow has been observed to be proportional to the drainage area in many regions of the world, area may then be used as a surrogate variable for discharge and $Q \bullet S$ = constant becomes $A^{0.5} \bullet S$ = constant. This relationship can be studied in detail using DEM's. The magnitude of a link n, is defined as the number of sources upstream of the link. For topological reasons the total number of links draining through the outlet of a link of magnitude n is $2n - 1$. The area draining directly to any link, A^*, varies randomly from link to link but does not depend on the magnitude. Thus the total area, A(n), draining through a link of magnitude n, is itself a random variable,

$$A(n) = \sum_{i=1}^{2n-1} A_i^* \tag{16}$$

Thus rather than to consider the energy expenditure per unit area of channel as a constant anywhere in the network, it is now considered as a random variable, ζ, whose expected value is the same throughout the neiwork. This is expressed as

$$Q^{0.5}(n)\ S(n) = \zeta(n) \tag{17}$$

where E $[\zeta(n)]$ = constant. Using A(n) as a surrogate of Q(n), Equation (17) yields in first order analysis

$$E[S(n)] = \text{constant} \bullet (2n - 1)^{-0.5} \tag{18}$$

Thus the principles of optimal energy expenditure lead to the scaling of the mean link slopes as function of areas or their surrogate variable magnitudes. The basic scaling is in terms of discharges but with discharges proportional to areas Equation (18) is a natural consequence. The scaling of Equation (18) with exponent of -0.5 is the one found in the analysis of drainage networks through DEM's by Tarboton, et al. (1989), an example of which is shown in Figure 4. From first order analysis of Equation (17) the variance of link slopes may be obtained as

$$\text{Var}[S(n)] \propto (2n - 1)^{-1}\ \text{Var}[\zeta(n)] \tag{19}$$

Flint (1974) and Tarboton et al. (1989) have found that Var[S(n)] scales as $(2n - 1)^{-0.5}$ which then implies that

$$\text{Var}[\zeta(n)] \propto (2n - 1)^{-0.5} \tag{20}$$

Figure 5 shows an example of the mean and variance of $A^{0.5}(n)\ S(n)$ as functions of magnitude for the Racoon River basin in Pennsylvania. Figure 5 is constructed grouping links according to magnitude so that there are at least 25 links with identifiable slopes in every interval of the histogram. Due to the 1m vertical resolution of DEM there are links whose altitude drop and the corresponding slopes cannot be identified. In these cases a random altitude drop between 0 and 1 meter is assigned to the link and the corresponding slope is then computed as the ratio

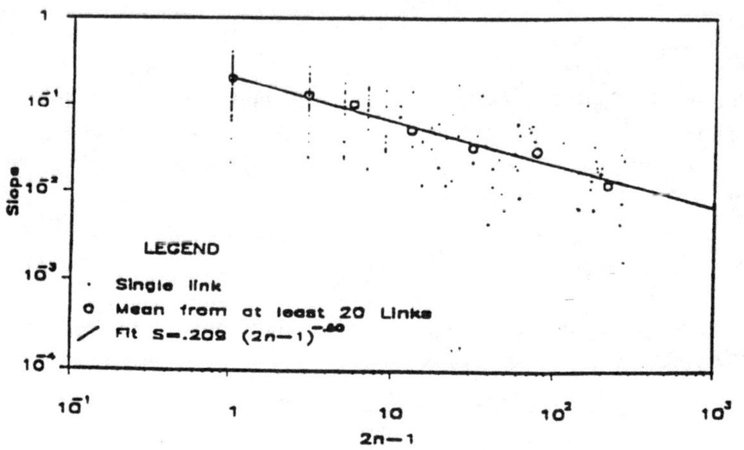

Figure 4. Link slopes as function of magnitude for Big Creek, Idaho [from Tarboton et al. (1991)].

of altitude drop and link length. The lines in Figure 5 were fitted by least squares through the whole set of points. It is observed that the mean is independent of magnitude, as predicted by the second principle of optimal energy expenditure, and variance is proportional to $(2n - 1)^{0.5}$. This is an indication of multiscaling in S(n) where there is not just one scaling relation determining all the moments of the process but rather that changes of scale affect different moments with different scaling laws.

The multiscaling character of S(n) points towards the fact that variance of the random variable whose expected value is the same throughout the network, increases proportionally to the average travel time for the flow to reach any site in the system. The reasons for this lie in the nature of the energy dissipation along any flow path in the network. Energy is spent along a succession of pools and riffles analogous to a diffusion of the energy along the flow path or equivalently to a random walk in the altitude space through which small drops of random height occur randomly along the flow path (Tarboton, et al.

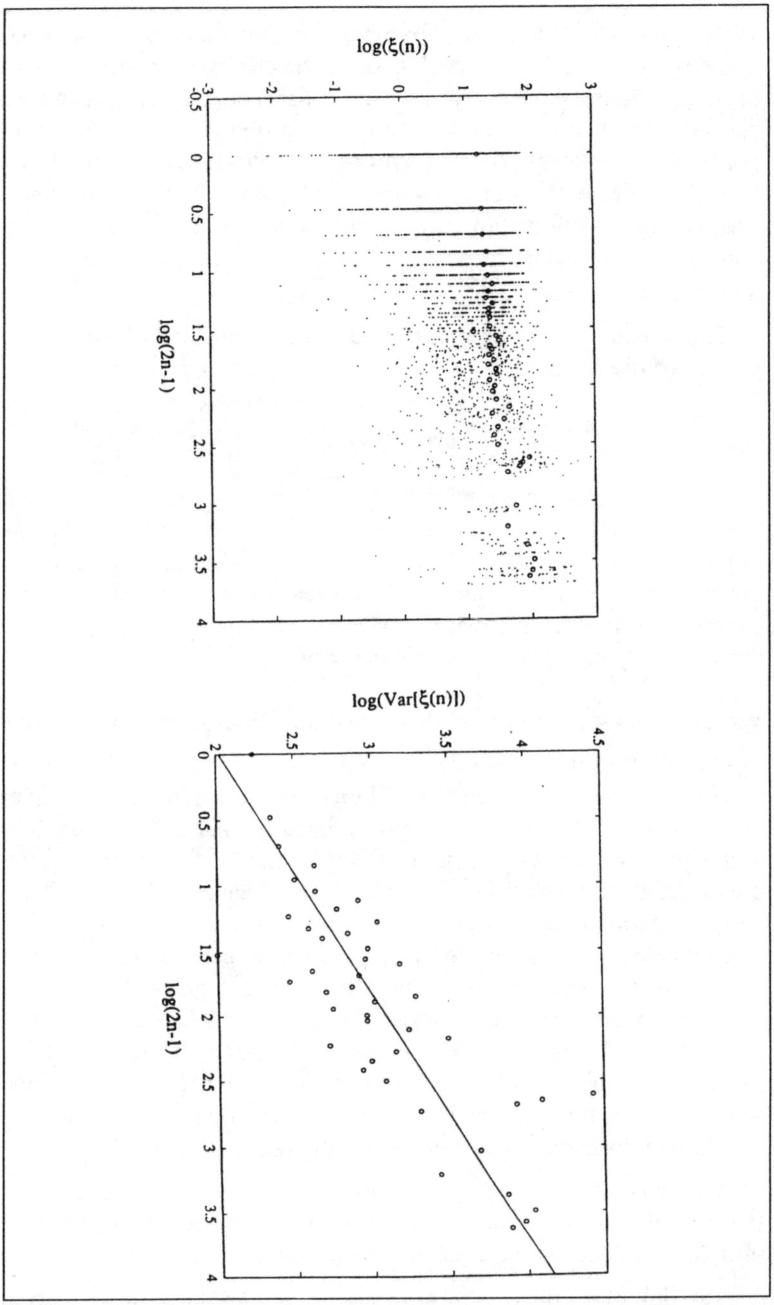

Figure 5. Mean and variance of ζ (n) vs. (2n – 1) for the Racoon River Basin in Pennsylvania.

1989). From the point of view of the optimal operation of the network, it is desirable that the expected value of the energy spent per unit area of channel be the same everywhere in the basin but the variance of such energy expenditure, similar to that in a diffusion process, will be proportional to the length of the path or equivalent to the average travel time for the flow to reach a particular link from all its tributary links. The average length of flow path at any point may be considered proportional to the square root of the total area draining at the point and thus $\mathrm{Var}[\zeta n)] \propto [(2n - 1)]^{0.5}$.

The spatial structure of runoff production is intimately linked to the scaling of the drainage network. Equation (17) may be written as

$$\left[i_1 \beta_1 A_1^* + 1_2 \beta_2 A_2^* + \ldots + i_{(2n-1)} \beta_{(2n-1)} A_{(2n-1)}^*\right]^{0.5} S(n) = \zeta(n) \quad (21)$$

where i is the mean annual rainfall input, and β is the mean annual runoff coefficient of the area draining into the individual links upstream of a link of magnitude n. The fact that $E[S(n)] \propto (2n - 1)^{0.5}$ and $E[\zeta(n)]$ = constant indicates that in first order analysis basins tend to be organized so that the expected value of annual runoff production per unit area, $i_1 \bullet \beta$, remains approximately the same throughout the basin.

Power-Law Distribution of Discharge and Energy in River Basins

One of the main obstacles in understanding surficial hydrologic processes is the high spatial variability of surface features in river basins to which those processes are intimately linked. River runoff is a key flux in climate systems. It occurs over a wide range of spatial scales: from the microscale of the individual channel link in a drainage network through the mesoscale of drainage basins, to the macroscale of continents. "The search for an invariance property across scales as a basic hidden order in hydrologic phenomena, to guide development of specific models and new efforts in measurements is one of the main themes of hydrologic science" (National Research Council, 1991). It has been the theme of this lecture that the existence of such hidden order in the spatial distribution of discharge and energy expenditures should be imprinted in the structure of the drainage network.

It is only recently with the availability of digital elevation maps (DEM's) that detailed and extensive study of the network at the level of links and contributing areas has been routinely possible.

Any link of a drainage network carries a wide range of discharges resulting from a variety of rainfall events and antecedent conditions of soil moisture. Nevertheless for our purposes it is assumed that the mean annual flow is a representative discharge of the conditions at every link. Flow measurements at every link of a network have not been collected and since the mean annual flow has been observed to be proportional to the drainage area in many regions of the world, as mentioned before in this lecture, the total cumulative area draining into a link will then be used as a surrogate variable for discharge. Five basins of very different characteristics throughout North America were used in the analysis. All of them have areas above 100 km^2 and drainage networks which defined through DEM's have more than 2000 links allowing a reliable probability distribution analysis of discharge and energy in the population of links. They are described in Table 1.

TABLE 1.
Characteristics of Basins Used in the DEM Analysis.

Name	Location	Area (km^2)	Dem Pixel Size (mxm)	Support Threshold in Pixels	Total Number of Links	Slope Power-Law Discharges	Slope Power Law Energy Releases
Brushy	AL	322	30 x 30	25	7251	-0.43	-0.91
Cald	ID	147	30 x 30	40	2187	-0.41	-0.93
Racoon	PA	445	30 x 30	50	5033	-0.42	-0.92
Scho	NY	2408	68.3 x 92.7	70	3089	-0.43	-0.90
St. Joe	MT, ID	2834	62.2 x 92.7	50	4997	-0.44	-0.92

Takayasu, et al. (1988) have shown that aggregation systems with constant injection of mass follow a power-law mass distribution. Their model is described as follows: particles with integer units of mass are placed on each site of a geometrical lattice and they aggregate by random walk processes with discrete time steps. Besides this aggregation process there is an injection effect, a particle with unit mass is added at every site at every time step. As a result there is no site without particle at any time step. An example of the structures developed by this model is shown in Figure 6. Although Figure 6 in Takayasu's interpretation has time as the vertical axis, the network can also be seen as a 2-D object, and it is equivalent to a stochastic model of river networks developed by Scheidegger (1967). After a long time the size distribution of this aggregation system with injection converges to a power-law size distribution independent of the initial conditions. The cumulative probability distribution of a site having mass larger than m satisfies

$$P[M > m] \propto m^{-\beta} \tag{22}$$

The value of the exponent depends on the characteristics of the random walk, β is 1/3 when at each time step particles will either

Figure 6. Dendritic structure formed by a Takayasu's aggregation model.

jump to the nearest neighbor site or stay at the same site with probability p = 1/2. With p = 1/2 the model yields dendritic structures similar to those found in river networks. Figure 6 corresponds to this case. The model operates in a one-dimensional lattice where each point has two neighbors; the vertical axis in Figure 6 represents time. At any constant in time one has then the connectivity relating each point to those in the one-dimensional lattice which up to that moment are linked to the point in consideration. As pointed out by Takayasu et al. (1988), this model is exactly equivalent to that of Scheidegger (1967) for drainage network patterns which operate in a two-dimensional space lattice rather than in time coupled with a one-dimensional lattice. A geometrical explanation of $\beta = 1/3$ is based on the fact that the mass distribution is equivalent to the distribution of the area draining to a point which in turn is the area surrounded by two random walk trajectories which would be the surrounding ridges in the boundary of the basin (Tayakasu, 1988). Dhar and Ramaswamy (1989) using a variant of the model of Bak et al. (1987, 1988) obtained similar results.

A river network under mean annual flow conditions may be considered an aggregation system with constant injection of particles and thus it seems reasonable to explore whether there is a power-law distribution for the mean annual discharge at any link or for its surrogate variable the cumulative area contributing to the link. Figure 7 shows this is indeed the case for the basins analyzed through DEM's. The distributions follow good straight lines for near 3 log-scales. The deviation at very large values of areas is likely to be a finite size effect, the number of links with such large area being very small and the statistics losing significance. Thus these deviations are expected to start earlier for those basins with fewer number of total links. The value of β is relatively unaffected by the size of the support threshold used in the identification of the network. Figure 8 shows the power-law distributions of mass as a function of the support threshold for one of the basins analyzed.

The reduction of the support threshold with the corresponding increase in the number of links does not change the slope but only tends to define it better over a longer range. Obviously there will be a limit after which one would not be working with the drainage network, but rather with the contributing hillslopes. It is important to notice that the exponent β is statistically undistinguishable among the different basins and approximately equal to 0.43. The apparent universality of the exponent points towards a self-organized critical phenomena as described by Bak et al. (1987, 1988,1990). In fact, Bak et al. (1987) have shown that some extended dissipative dynamical systems naturally evolve into a critical state with no characteristic time or length scale, their spatial signature is the emergence of a scale invariant (fractal) structure which becomes apparent through a power-law distribution. This will be discussed later in this lecture. An explanation of the value $\beta = 0.43$ based on the fractal nature of rivers is now offered.

A well known empirical relationship based on many basins of widely different sizes is $L \propto A^{0.568}$ which provides a very good fit between the length L of the main stream and the area of the basin, A (Gray, 1961). Taking the exponent as 0.6, Mandelbrot (1983) suggested that rivers are fractals with fractal dimension $D = 1.2$. A better estimate is $D = 2 \times 0.568 \simeq 1.1$ which was corroborated by Tarboton et al. (1988) in their analysis of DEM's using Richardson's divider method and functional box counting. The dimension $D \simeq 1.1$ is due to the sinuosity of individual rivers and corresponds to the scaling implied by $L \propto {}^{0.568}$.

Following Mandelbrot (1983) one may think of the ridges defining the boundaries of the contributing area to any link as mirror images of the rivers themselves and with the same fractal dimension. The Euclidean length L_e from the outlet to the most distant point in the

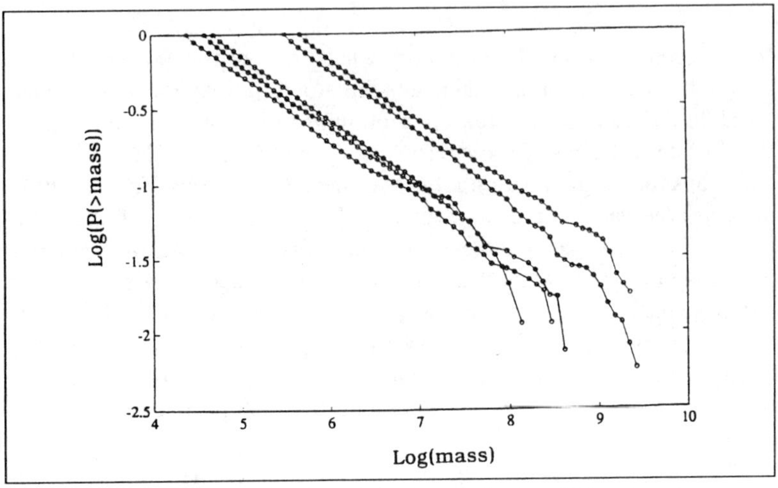

Figure 7. Cumulative distribution of discharges in five different basins.

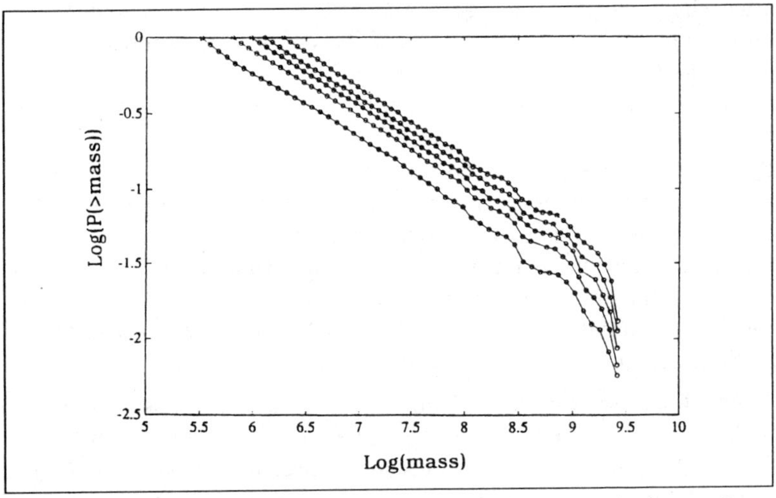

Figure 8. Cumulative distribution of discharges for the same basin as a function of the support threshold.

boundary can then be regarded as the first collision time of two fractal trails. As pointed out by Takayasu et al. (1988) in the case of Brownian motion, the distribution of L_e is approximately given by the well known recurrence time distribution of Brownian motion in one dimension.

$$p(L_e) \simeq (2\pi)^{-1/2} e^{-1(2L_e)} L_e^{-3/2} \propto L_e^{-3/2} \tag{23}$$

As compared with Brownian motion, fractal motion moves anomalously large distances from the origin. This can be accounted through a fractal diffusivity defined as

$$D_H = D\,|L_e|^{2H-1} \tag{24}$$

(Feder, 1988). This in turn implies for a river basin which reduces to Equation (23) when H = 0.5. The value of H is directly related to the

$$p(L_e) \propto (L_e^3 L_e^{2H-1})^{-1/2} \tag{25}$$

fractal dimension and for the case of a divider type of dimension, H = I/D (Feder, 1988). With D = 1.1, Equation (25) yields

$$p(L_e) \propto L_e^{-1.909} \tag{26}$$

Since $L_e \propto A^{0.5}$ one gets,

$$p(A>a) = p(L_e > a^{0.5}) \tag{27}$$

and thus,

$$p(A > a) \propto \int_{a_{0.5}}^{\infty} L_e^{-1.909} dL_e = a^{-0.45} \tag{28}$$

which agrees well with the relationship empirically found in Figure 7.

Mandelbrot (1974) has shown that uniform energy input in extended dissipative systems frequently results in power-law spatial distributions of energy storage and fractal energy dissipation. Bak et al. (1987, 1988, 1990) have interpreted this as a manifestation of a critical state, which is an attractor for the dynamics being robust with respect to variations in the parameters, the initial conditions and the presence of randomness. Thus dynamical systems with extended spatial degrees of freedom in 2 or 3 dimensions naturally evolve in self-organized critical states on which the energy is dissipated at all length scales. It is essential that the systems are dissipative (energy is released), that they are spatially extended with an "infinity" of degrees of freedom, and that energy is fed into the system continuously.

River networks are precisely this kind of system where the different channel links and their contributing areas support each other in a way which cannot be understood by studying the individual components in isolation and where the holistic view offered by self-organized criticality may be of great use. Thus one would expect river networks to exhibit a power-law spatial distribution of energy dissipation similar to the one found in other dissipative phenomena like earthquakes. For example, the well known Gutenberg-Richter law which gives a number of earthquakes with magnitude greater than a certain value has been shown to be equivalent to a power law of the energy released by the events (Aki, 1981; Turcotte, 1989; and Bak and Tang, 1989). The rate of energy expenditure per unit length of channel at any link of a river network is proportional to the product of discharge and slope, Q • S. Figure 9 shows the cumulative distributions of energy release through the links of the different networks of Table 1 where again area has been used as a surrogate variable of discharge. The oscillations for large values of energy expenditure are likely to be due to the small number of links in this region which makes the statistics unreliable. The flattening of the distribution for small energy releases is similar to that observed in earthquakes (Bak and Chen, 1991) and may be due either to problems of resolution in case of small slopes or to the fact that power-law behavior takes place after a certain threshold. This is not crucial since the region of interest is the right-hand side of the distribution. The power-law exponent is in all cases close to 0.90 which will be now related to the exponent 0.45 found for the mass distribution.

I have argued previously in this lecture that the rate of energy expenditure per unit length of channel at a link transporting a discharge Q may be expressed as $KQ^{1/2}$ where for a given flow condition, e.g., mean annual flow throughout the basin, K is a constant throughout the network. Thus the distribution of energy release is proportion to $Q^{1/2}$ and using area as a surrogate variable of discharge one gets

$$P[E > e] \propto P\left[A^{1/2} > a\right] \propto \int_{a_2}^{\infty} x^{-1.45}\,dx = a^{-0.90} \qquad (29)$$

which matches the empirical analysis and again tends to validate the principle of energy expenditure in river networks previously mentioned.

Final Comments

The main theme of this Tenth Kisiel Lecture has been my belief that behind the many observed empirical relations among the parts of a

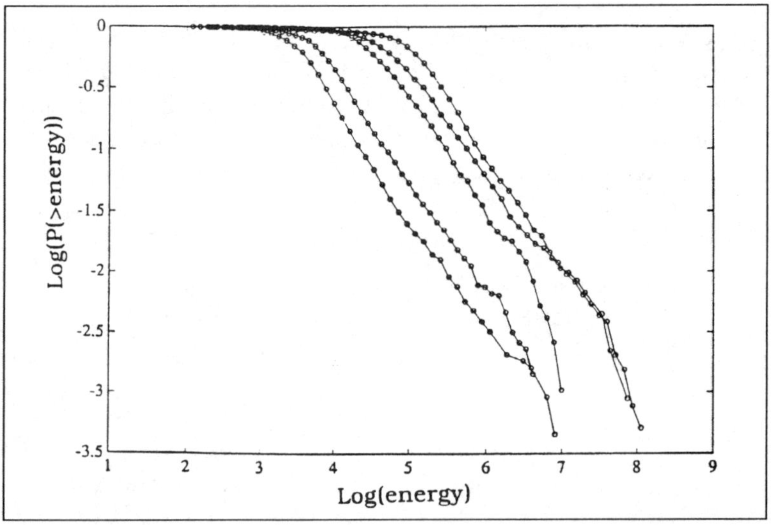

Figure 9. Cumulative distribution of energy release in five different basins.

drainage network there are some basic unifying principles rooted in the optimum energy expenditure framework. Moreover not only will these principles provide an explanation for those empirical relations and will link them to the process of runoff production, furthermore they will constitute in my opinion the element of "necessity" that has been missing in this topic, where most of the findings have been related only to the "chance" aspects of the processes. Quite exciting also is the fact that the power-laws describing the behavior of the key hydrologic variables have been found to hold with the same exponent over very different basins and that these laws are also reproduced under the principles of optimal energy expenditure. Since power-laws are the signature of fractal type of structures it would appear that the fractal characteristics of river networks are a consequence of energy type for considerations. If correct, this will be an important finding whose implications would extend beyond hydrology.

Acknowledgements

All the results and ideas of this lecture are from a close collaborative effort involving researchers in Italy, U.S.A. and Venezuela. I want to thank Professors Andrea Rinaldo, Rafael L. Bras, Alessandro Marani and David Tarboton, as well as Mr. Riccardo Rigon and Mr. Ede Ijjász-Vásquez not only for their efforts and ideas but also for an exciting and most enjoyable research atmosphere which has been kept via telephone, fax and express-mail. The lecture itself is taken from the papers entitled (1) Energy Dissipation, Runoff Production, and the 3-Dimensional Structure of River Basins; (2) Power-Law Distribution of Discharge and Energy in River Basins, and (3) Minimum Energy and Fractal Structures of Drainage Networks. All of these papers will be published in Water Resources Research.

REFERENCES

Aki, K., A probabilistic synthesis of precursory phenomena, in *Earthquake Prediction,* (D.W. Simpson and P.G. Richards, eds.) American Geophysical Union, Washington, DC, pp. 556-574, 1981.

Bak, P. and K. Chen, Self-organized criticality, *Scientific American,* 265: 46-53, 1991.

Bak, P., K. Chen, and C. Tang, A forest-fire model and some thoughts on turbulence, *Physical Review Letter. A.,* 147:297-300, 1990.

Bak, P. and C. Tang, Earthquakes as a self-organized critical phenomenon, *Journal of Geophysical Research,* 94:15635-15637, 1989.

Bak, P., C. Tang and K. Wiesenfeld, Self-organized criticality: An explanation of 1/f noise, *Physical Review Letter,* 59:381-384, 1987.

Bak, P., C. Tang and K. Wiesenfeld, Self-organized criticality, *Physical Review A.,* 38:364-374, 1988.

Band, L.E., Topographic partition of watersheds with digital elevation models, *Water Resources Research,* 22(1): 15-24, 1986.

Brush, L.M., Jr., Drainage basins, channels and flow characteristics of selected streams in central Pennsylvania, U.S. Geol. Surv. Prof. Paper 282-F, 1961.

Dhar, D. and R. Ramaswamy, Exactly solved model of self-organized critical phenomena, *Physical Review Letter,* 63:1659-1662, 1989.

Feder, J., *Fractals,* Plenum, New York, 1988.

Flint, J.J., Stream Gradient as a function of order, magnitude and discharge, *Water Resources Research,* 10(5):969-973, 1974.

Gray, D.M., Interrelationships of watershed characteristics, *Journal of Geophysical Research,* 66:1215-1223,1961.

Horton, R.E., Erosional development of streams and their drainage basins; hydrophysical approach to quantitative morphology, *Bulletin of the Geological Society of America,* 56:275-370, 1945.

Howard, A.D., Simulation of stream networks and headward growth and branching, *Geographical Analysis,* 3:29-50, 1971.

Howard, A.D., Theoretical model of optimal drainage networks, *Water Resources Research,* 26(9): 2107-2117, 1990.

Ikeda, S., G. Parker, and Y. Kimura, Stable width and depth of straight gravel rivers with heterogeneous bed materials, *Water Resources Research,* 25:713-722,1988.

LaBarbera, P. and R. Rosso, The fractal geometry of river networks, *EOS,* Trans. Amer. Geoph. Union, 68:1276, 1987.

Leopold, L.B., and W.L. Langbein, The concept of entropy in landscape evolution, U.S.G.S. Professional Paper 500-A, 1962.

Leopold, L.B., and T. Maddock, Jr., The hydraulic geometry of stream channels and some physiographic implications, U.S. Geol. Surv. Prof. paper 252, 1953.

Leopold, L.B. and M.G. Wolman, River channel patterns: braided, meandering, and straight, U.S. Geol. Surv. Prof. paper 282-B, 1957.

Leopold, L.B., M.G. Wolman, and J.P. Miller, *Fluvial Process in Geomorphology,* W.H. Freeman and Co., San Francisco, 1964.

Lin, S., Computer solutions for the traveling salesman problem, *Bell Syst. Tech. Jour.,* 44:2245-2269, 1965.

Mandelbrot, B.B., Intermittent turbulence in self-similar cascades: Divergence of high moments and dimensions of the carrier, *Journal of Fluid Mechanics,* 62:331-358, 1974.

Mandelbrot, B.B., *The Fractal Geometry of Nature,* W.H. Freeman, New York, 1983. National Research Council, Committee on Opportunities in the Hydrologic Science, *Opportunities in the Hydrologic Sciences,* National Academy Press, Washington, 1991.

Murray, C.D., The physiological principle of minimum work: I. The vascular system and the cost of blood volume, *Proc. Natl. Acad. Sci.,* 12:207-214, 1926.

O'Callaghan, J.F. and D.M. Marks, The extraction of drainage networks from digital elevation data, *Computer Vision Graphics and Image Processing,* 28:323-344, 1984.

Parker, G., Self-formed straight rivers with equilibrium banks and mobile bed. Part 2. The gravel river, *Journal of Fluid Mechanics,* 89:127-146, 1978.

Pilgrim, P.H., Isochrones of travel time and distribution of flood storage from a tracer study on a small watershed, *Water Resources Research,* 13(3):587-595, 1977.

Rinaldo, A., I. Rodríguez-Iturbe, R. Rigon, R.L. Bras, E. Ijjaz-Vasquez and A. Marani, Minimum energy and fractal structures of drainage networks, submitted to *Water Resources Research,* 1991.

Rodríguez-Iturbe, I., E.l. Ijjasz-Vasquez, R.L. Bras, and D.G. Tarboton, Power-Law Distributions of Discharge and Energy in River Basins, submitted for publication to *Water Resources Research,* 1991.

Roy, A.G., Optimal angular geometry models for river branching, *Geographical Analysis,* 15:87-96, 1983.

Scheidegger, A.E., A stochastic model for drainage patterns into an intramontane trench, *Bulletin Association Scientific Hydrology,* 12:15-20, 1967.

Sherman, T.F., On connecting large vessels to small, *J. Gen. Physiol.,* 78(4):431-453, 1981.

Shreve, R.L., Statistical law of stream numbers, *Journal of Geology,* 74:17-37, 1966.

Shreve, R.L., Stream length and basin areas in topologically random channel networks, *Journal of Geology,* 77:397-414, 1969.

Stevens, P.S., *Patterns in Nature,* Little, Brown and Co., Boston, 1974.

Takayasu, H., I Nishikawa, and H. Tasaki, Power-law mass distribution of aggregation systems with injection, *Phys. Review A.,* 37:3110-3117,1988.

Tarboton, D.G., R.L. Bras, and I. Rodríguez-Iturbe, The fractal nature of river networks, *Water Resources Research,* 24(8): 1317-1322, 1988.

Tarboton, D.G., R.L. Bras, and I. Rodríguez-Iturbe, The analysis of river basins and channel networks using digital terrain data, M.I.T., Ralph Parsons Laboratory Report No. 326, 1989.

Tarboton, D.G., R.L. Bras, and I. Rodríguez-Iturbe, Scaling and elevation in river networks, *Water Resources Research,* 25(9):2037-2051, 1989.

Tarboton, D.G., R.L. Bras, and I. Rodríguez-Iturbe, Scaling and drainage density in river basins, *Review of Geophysics,* submitted for publication, 1991.

Turcotte, D.L., Earthquake prediction, in *Fractals in Geophysics,* (C.H. Scholz and B.B. Mandelbrot, eds.), Birkhausen, Basel, Switzerland, pp. 171-196, 1989.

Uylings, H.B.M., Optimization of diameters and bifurcation angles in lung and vascular tree structures, *Bull. Math. Biol.,* 39:509-520, 1977.

Yang, C.T., Potential energy and stream morphology, *Water Resources Research,* 7(2):311- 322,1971.

Woldenberg, M.J. and K. Horsfield, Relation of branching angles to optimality for four cost principles, *Journal of Theoretical Biology,* 122:204, 1986.

Wolman, M.G., The natural channel of Brandywine Creek, Pennsylvania, U.S., U.S.G.S. Professional Paper 271, 56 pp., 1955.

BIOGRAPHY

Ignacío Rodríguez-Iturbe

Ignacío Rodríguez-Iturbe was born in Caracas, Venezuela in 1942. His primary high school and university education took place in Maracaibo, where he graduated with the prize "Máxima Calificación" in the Civil Engineering School of the Universidad del Zulia in 1963. After working for one year at Zulia he entered Catlech, where he obtained his M.Sc. in 1965 and then Colorado State University, where he received his Ph.D. in 1967. From 1967 to 1969 he taught at Zulia University and from 1969 to 1971 he was at the Instituto Venezolano de Investigaciones Científicas. From 1971 to 1975 he served as Associate Professor and the Associate Head in the Water Resources Division (Ralph Parsons Laboratory) of the Civil Engineering Department of the Massachusetts Institute of Technology. From 1975 to 1987 he was associated with Universidad Simon Bolívar as Professor of Engineering. At USB he also served as Head of the Graduate Program in Hydrology (which he founded), Dean of the Graduate School and Dean for Research. From 1987 to the present he has been Professor of Hydrology at the Instituto Internacional de Estudios Avanzados (I.D.E.A.) in Caracas. In the period 1989-1991 he was the Endowed Professor of Civil and Environmental Engineering at the Institute for Hydraulic Research of the University of Iowa. Throughout the years he has been associated with M.I.T., where he holds a joint appointment as Senior Lecturer or Visiting Professor.

Ignacío Rodríguez-Iturbe's major scientific and professional interest is in theoretical hydrology. He is the author of a text in Random Functions and Hydrology (with Rafael L. Bras) and editor of several monographs in different aspects of hydrology. His honors and awards include the Robert Horton Award (1975) and James B. Macelwane Award (1977) of the American Geophysical Union; the Huber Research Prize (1975) of the A.S.C.E., the Francisco Torrealba Research Prize of Universidad Simón Bolívar (1985), the Antonio Borjas Distinguished Alumni Medal of Universidad del Zulia (1991), the Academic Medal of the University of Florence (1991) and the 1987 Prize for the best research paper in Engineering from the Venezuelan National Prize and in 1992 an Honorary Doctor's Degree from the University of Genoa. Ignacio Rodrfguez-Iturbe is a member of the Latin American Academy of Sciences (1983), the Third World Academy of Sciences (1988) and in 1988 he was elected a Foreign Associate of the National Academy of Engineering.

FOREWORD

In the foreword of the Fourth Kisiel Memorial Lecture[1], we have suggested that water resources, as a scientific endeavor, could be defined as the study of man's intervention in the hydrological cycle. Humans, indeed, intervene in nature's hydrological cycle in a variety of ways, affecting, among others hydrogeochemical as well as hydrogeobiological balances. The outcome of these disturbed balances appears often as the degraded quality of the aquatic environment.

Ever since the human species inhabited planet Earth, it joined the other species on it in discharging into the environment the waste products generated by the metabolic activity. During the last half century, however, the nature of the pollutants has changed significantly: from a mostly catabolic character (e.g., excreta of humans and of other species), to discarded wastes of the modern, industrialized society (e.g., synthetic organics, by-products of fuel combustion, radio-active materials). The discharge into the environment of the modern contaminants are phenomena that we observe at the local or regional scale. These processes, however, may influence significantly major hydrogeochemical cycles.

The understanding of anthropogenic influences on the quality of the environment requires proper re-orientation of classical scientific disciplines. For example, since many insults to the environment affect directly microflora and microfauna, the classical microbiology needs to be re-focused as environmental microbiology, departing from the paradigm "one disease - one microorganism" and establishing a new point of view that considers diverse populations of microorganisms, each group performing a specific task in the environment. Similarly, classical toxicology which deals with the effect of toxic substances on living organisms needs to branch out into ecotoxicology, emphasizing public health effects of toxics discharged into the environment. A further stage of scientific inquiry might study the synergetic effects on human well-being of pathogens and toxics. In addition, these studies can contribute significantly to our understanding of global effects of local and regional ecological deterioration.

Nathan Buras

---

1 Peter S. Eagleson, *The Emergence of Global-Scale Hydrology*, March 20, 1995.

SURFACE CHEMICAL THEORY AND PREDICTING THE DISTRIBUTION OF CONTAMINANTS IN THE AQUATIC ENVIRONMENT

by Werner Stumm
Swiss Federal Institute of Technology, ETH Zurich
Institute for Water Resources and
Water Pollution Control (EAWAG)
CH-8600 Dubendorf, Switzerland

Introduction

The various interconnected reservoirs of the Earth (atmosphere, water, soil, sediments, biota) contain material that is characterized by high area-to-volume ratios. A highly efficient interface chemistry assists in maintaining resilience in our environment and regulates water composition. Humans intervene in the hydrological cycle. Adsorption influences the distribution of pollutants between the aqueous phase and particulate matter and, in turn, affects their transport through the various reservoirs of the Earth. The geochemical fate, the residence time, and the residual concentrations of reactive elements (heavy metals) and organic pollutants are to a large extent controlled by their affinity to the solid surface. Colloids are ubiquitous in seawater, in fresh surface waters, in soil, and in groundwaters; a renewed interest concerns the role of colloids in the transport of reactive elements, radionuclides, and other pollutants.

The chemical, physical, and biological processes that occur at the mineral-water, the particle-water, and the organism-water interface influence the major geochemical cycles. The hydrologist is needed to participate in a multidisciplinary partnership in the assessment of the interaction of the hydrological cycle with the other biogeochemical cycles. Understanding how chemical cycles interdepend and are coupled by particles and organisms may aid our understanding of global ecosystems and how interacting systems may become disturbed by civilization.

Antropogenic Interference in Hydrogeochemical Cycles

The water cycle is intimately connected to biogeochemical cycles (Fig. 1). The definition on hydrology states (Buras, 1985) that it is the science which studies the interaction on the planet Earth between the hydrosphere on the one hand and the lithosphere and biosphere on the

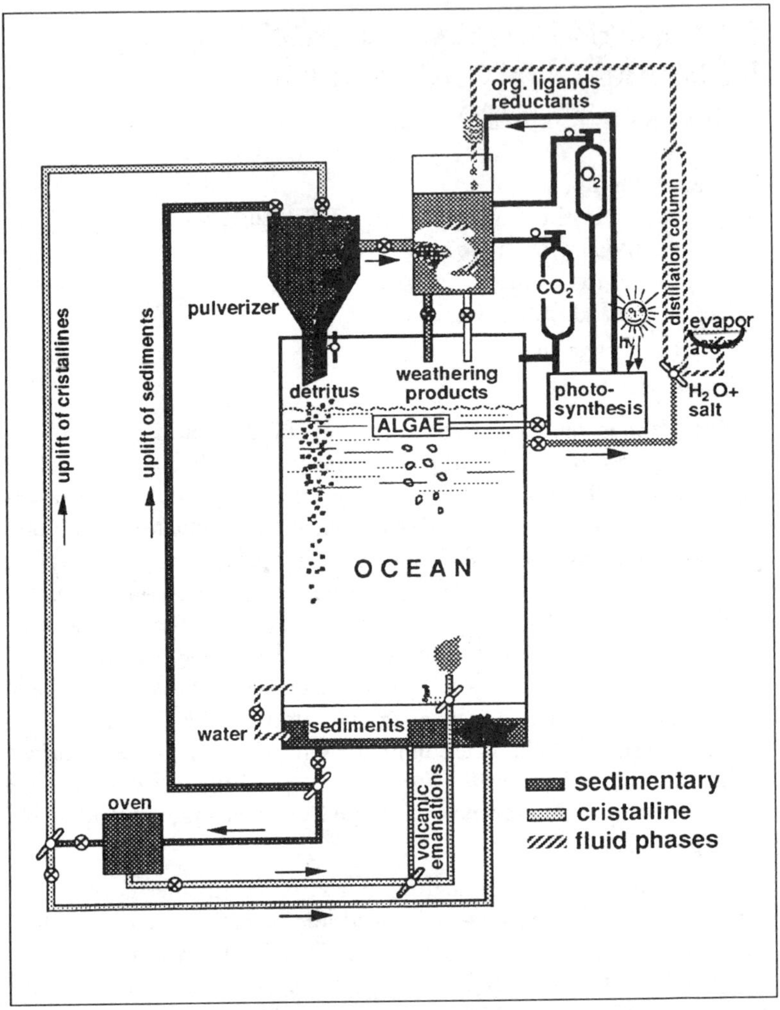

Figure 1. Circulation of rocks, water and biota. Steady-state model for the Earth's surface geochemical system likened to a chemical engineering plant. The interaction of water with rocks in the presence of photosynthesized organic matter continuously produces reactive material of high surface area in the surface environment. This process provides nutrient supply to the biosphere and, along with biota, forms the array of small particles (soils). Weathering imparts solutes to the water and erosion brings particles into surface waters and oceans. A large flux of settling detrital and biogenic particles continuously runs through the water column. The steady-state conveyor belt of settling particles which are efficient sorbents of heavy metals and other trace elements regulates their concentrations in the water column. The sediments are the predominant sink of trace elements.

other. As pointed out in the 1985 Kisiel lecture human intervention in the hydrological cycle is relevant to hydrology. Humans have become geochemical manipulators and agents of global change and are a major force in the transport of solid Earth materials. Chemical byproducts are changing the hydrosphere and atmosphere. The change on our planet involves a complex interaction between the inorganic physical processes and the biological processes. Although we depend on human intervention in the water cycle to provide adequate supplies for agriculture and people, we are often not sufficiently aware of some of the negative effects that a mammoth water diversion may have on ecosystems. The consequences may not be apparent for some time, and causes and effects may be difficult to identify due to the large distances involved. The weathering cycle is affected markedly at least locally and regionally by our civilization. In local environments, proton (H^+) and electron (e^-) balances may become upset, and significant variations in pH and redox intensity (pE) occur.

The Sensitivity of Fresh Waters to Perturbation

In Figure 2, the sizes of the various reservoirs, measured in number of molecules of atoms, are compared. The mean residence time of the molecules in these reservoirs is also indicated. The smaller the relative reservoir size and residence time, the more sensitive is the reservoir toward perturbation. Obviously, the atmosphere, living biomass (mostly forests), and ground and surface fresh waters are most sensitive to perturbation. The anthropogenic exploitation of the larger sedimentary organic carbon reservoir (fossil fuels and byproducts of their combustion such as oxides and heavy metals and the synthetic chemicals derived from organic carbon) can, above all, affect the small reservoirs. Over the past years, we have started to recognize that biosphere processes play an important role in coupling the cycles of essential elements and in regulating the chemistry and physics of our environment. The living biomass (Fig. 2) is a relatively small reservoir and is subject to anthropogenic interference; each species forming the biosphere requires specific environmental conditions for sustenance and survival.

Chemical Dynamics of Pollutants

Let us consider the release of a potential pollutant into the environment. The transfer of the pollutant into various reservoirs (air, water, soil, biota, etc.) and its ultimate distribution and residual concentration (activity or fugacity) (Fig. 3) depend a) on the physical, chemical and biochemical (compound-specific) *properties of the pollutant* (vapor pressure, solubility, functional groups, Henry coefficient, lipophility, adsorbability, chemical and biological degradability), and b) on the

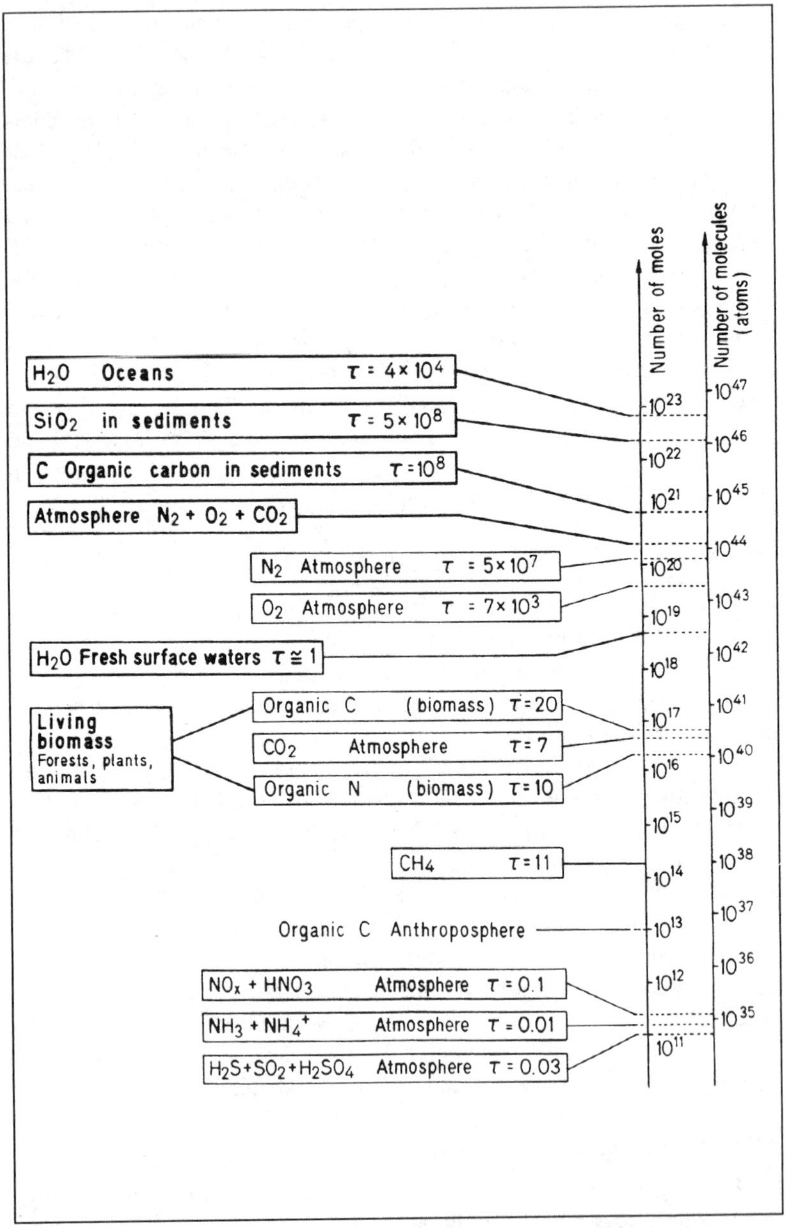

Figure 2. Comparison of global reservoirs. The reservoirs of atmosphere, surface fresh waters and living biomass are significantly smaller than the reservoirs of sediment and marine waters. The total groundwater reservoir may be twice that of fresh surface water. However, groundwater is much less accessible (r = respective residence time [years] of molecule [atoms]) (Stumm, 1986).

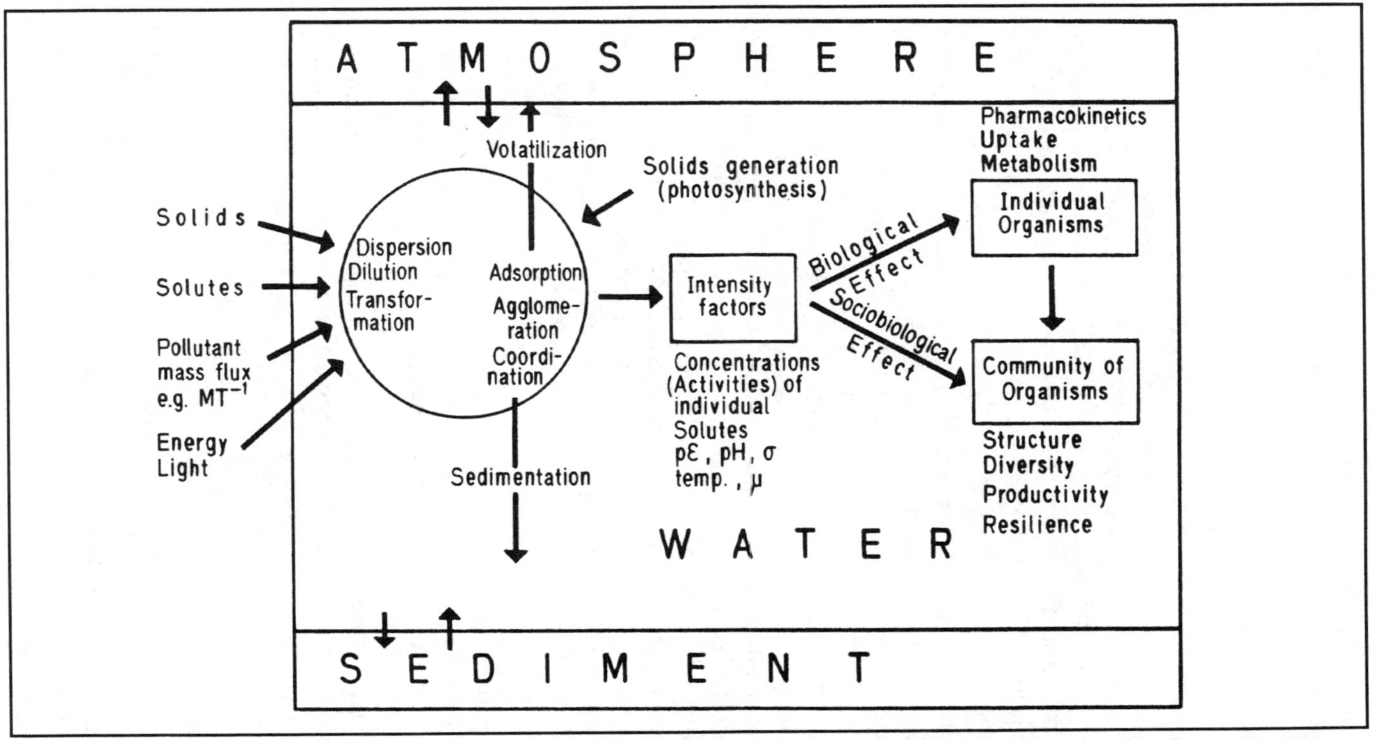

Figure 3. Transfer and transformation of pollutants in aquatic ecosystems by various chemical, biological and physical processes and pathways. A substance introduced into the system becomes dispersed or diluted. It can become eliminated from the water by adsorption on particles or by volatilization. It may also be transformed chemically or biologically.

properties of the environment (flow of water, type of interfaces, transport of particles, photosynthesis and nutrient cycles, redox intensity, presence of other constituents, etc.) (Table 1).

The character of pollution has changed over the last decades especially in industrialized and densely populated regions. Over the years, the gross pollutional load has increased and its character has changed. While a few decades ago, most of our wastes were predominantly catabolic (excreta of man and animals and other biogenic components), they are now more and more composed of discarded material from modern industrial society (synthetic chemicals, mining products [phosphates, metals], byproducts of fossil fuel combustion and energy production [metals, oxides of S, N, H, heat, radioactive isotopes]).

Many of the industrial chemicals reach receiving waters indirectly (via households, agricultural drainage, atmosphere). The changes in the character of pollutional load become especially apparent in areas of high population and industrial density.

Monitoring data rarely can be generalized unless one knows the significant connection and interactions between the parts of an ecological system. Similarly, it is usually not possible to predict the effects of chemicals on ecology and human health either from bioassays or from tests with individual organisms under laboratory conditions. In view of the great number of existing industrial chemicals, and considering the large amount of environmental pollutants created every day by human activities, we need to develop and apply general concepts on the behavior and fate of pollutants and to use these concepts as a means to procure the relevant analytical results.

By evaluating the strength of natural and anthropogenic emission sources and by identifying the relevant pathways and describing the interactions and the unit processes that govern the behavior and fate of chemicals in a given natural system, we can improve our ability to understand and predict the future fluxes and distribution of pollutants and their effects on ecological systems and human health. Table 1 identifies some of the relevant parameters and unit processes that govern the spatial and temporal distribution of pollutants.

Residence time. The residence time in a body of water is influenced by hydrological factors (type of dispersion and loss through the outlet) and by non-hydrological processes (e.g. adsorption on settling particles or biomass and chemical and biological transformations). Among the molecular transformations, *biodegradability* is one of the most important factors determining the residence time of organic pollutants in the receiving waters. Today, detailed knowledge of the microbiology, the

TABLE I.
Pertinent Information to assess Fate and Residence Time of Pollutants

	PARAMETER	SIGNIFICANCE
Transport Paths Mass Flows	production statistics	immision into environment
	mass balance	
DISTRIBUTION air-water surface-water sediment-water	adsorption isotherms	
		sedimentation
	solubility	
	vapor pressure	immission into and from atmosphere
	Henry constant	evaporation condensation
	gas transfer coefficient	
Biomagnification	lipophility	accumulation food chain
	n-octanol-water distribution coeff.	biolog. retention
	solubility	
Molecular Transformation (**microbial oxidation hydrolysis photolysis**)	biodegradability	residence time half life
	equilibrium constants	
		structure-reactivity correlations
	rate constants	
		photodecomposition
	light absorption	

mechanisms, and the kinetics of the biotransformation of organic pollutants under natural conditions is still missing. Kinetic rate constants for *chemical transformations* (hydrolysis, photolysis, redox reactions), however, may often be predicted from *structure reactivity correlations.* The basis of such correlations is the well-known fact that, despite great diversity in structure, most organic chemicals share common reaction patterns and reactivities among like-structured chemicals.

Biomagnification. Of particular importance in assessing ecological or toxicological relevance of a pollutant is the question whether the substance is incorporated into the biota. Such incorporation may lead to a retention of the pollutant and its biomagnification in the food chain.

Incorporation into biomass is related to the lipophility as measured by the n-octanol water distribution coefficient. Because the latter is inversely related to solubility, biomagnification is usually inversely proportional to solubility.

Ecological or Toxicological Impact

For an assessment of exposure and an evaluation of the ecological and hygienic risk, we need to know the residual concentrations of pollutants. An estimate of the ambient concentration of a pollutant based on a prediction of its relevant fate and residence time (considering relevant pathways, exchange processes, biological and chemical conversions) or on the basis of analytical determinations in the receiving waters is essential to any hazard assessment. The problem of estimating the dose (activity) is one of the things that distinguishes ecotoxicology from classical toxicology.

The ecological harmfulness of a substance depends on its interaction with organisms or with communities of organisms (Fig. 3). The intensity of this interaction depends on the specific structure and activity of the substance under considerations but other factors such as temperature, turbulence, and the presence of other substances are also important. Comparative toxicological research is necessary in order to extrapolate effects observed in laboratory experiments to organisms in nature, from one organism to another or to humans.

In evaluating toxicity, we need to distinguish between substances that have a direct effect on humans, animals, and other terrestrial or aquatic organisms and substances that primarily affect the organization and structure of an ecosystem. Here a contaminant may impair the self-regulatory functions of the system or subtly interfere with food chains.

While we have some knowledge about the impact of xenobiotic substances on individual organisms, we know less about their impact on ecosystems. In considering biological communities, various intra- and inter-species interactions of a sociobiological nature need to be taken into account.

The natural distribution of organisms depends primarily on their ability to compete under given conditions and not merely on their ability to survive in the physical and chemical environment. A population will be eliminated when its competitive power is reduced to such an extent that it can be replaced by another species. The competitive abilities of an organism are based on its reproductive rate (which is related to food and physiological potential) and the mortality rate from all causes, including predation and imposed toxicity. There are many ways in which an organism can die, but there is only a very narrow range of ways in which it can survive and leave offsprings. Thus, in an ecosystem, a population may be eliminated by the presence of pollutants even at apparently trivial toxicity levels, if its competitive ability is marginal or if it is the most sensitive of the competitors.

Often, contaminants at very low concentrations cause changes in the structure of the population by interfering through chemotaxis with interorganismic communication. For example, the survival of a fish population may be rendered impossible by a pollutant (even if it exhibits neither acute nor chronic toxicity to the particular species of fish) if it impairs the food source (zooplankton) or disturbs chemotactical stimuli or mimics wrong signals (and thus, for example, interferes with food finding).

As a consequence, of the many microhabitats (niches) that are typically present in a healthy water body, many species can survive. Because of interspecies competition, most species are present in a low population density. Pollution destroys microhabitats, diminishes the chance of survival for some of the species, and thus, in turn, reduces the competition; the more tolerant species become more numerous.

This shift in the frequency distribution of the species toward a lower diversity of the ecosystem is a general consequence of the chemical impact on waters by substances not indigenous to nature.

An understanding of the interaction of chemical compounds in the natural system hinges on the recognition of the compositional complexity of the environment. This requires an adequate analytical methodology, especially the ability to detect individual components (chemical species) selectively and to measure them accurately and with sensitivity.

TABLE II.

Adsorption Equilibria as Defined: (A) for Surface complex Formation

(A) Surface Complex Formation Example: Binding of Cu(II) to a surface SH:

$$SH + Cu^{2+} \rightleftharpoons SCu^{+} + H^{+} \tag{1}$$

Applying the mass law to Eq. (1)

$$\frac{[SCu^{+}]}{[Cu^{2+}][SH]} = K^{s}_{app}[H^{+}]^{-1} \tag{2a}$$

K^{s}_{app} depends on the surface charge or surface potential. This can be corrected (Stumm, 1992) by

$$K^{s}_{app} = K^{s} \exp\left(-\frac{F\psi}{RT}\right) \text{ Thus, Eq. 2a) becomes}$$

$$\frac{[SCu^{+}]}{[Cu^{2+}][SH]} = K^{s}[H^{+}]^{-1} \exp\left(-\frac{F\psi}{RT}\right) \tag{2b}$$

where:
F = Faraday
ψ = surface potential
R = gas constant
K^{s} is an intrinsic constant, valid for a non-charged surface

ψ cannot be determined but it can be calculated from the surface charge, *a,* (which is experimentally accessible) by the following simplified equation (valid for 25 ^{0}C; *a* is expressed in Coulomb m-^{2}, C in mol c, and T in volt).

$$\sigma = 0.1174C^{1/2} \sinh(z\psi \times 19.46) \tag{3}$$

Note that Eq. (2a) can be converted into a Langmuir type of equation.

$$\Gamma_{Cu} = \frac{\Gamma_{max} K^{a}_{app}[H^{+}]^{-1}[Cu^{2+}]}{1 + K^{s}_{app}[H^{+}]^{-1}[Cu^{2+}]} \tag{4}$$

where

Γ = $[SCu^{+}]$ / mass adsorbent
Γ_{max} = S_T / mass adsorbent
S_T = $[SCu^{+}] + [SH]$

Thus, at a given pH

$$\Gamma_{Cu} = \frac{\Gamma_{max} K^{s}_{ads(pH)}[Cu^{2+}]}{1 + K^{s}_{ads(pH)}[Cu^{2+}]} \tag{5}$$

where

$K_{ads(pH)} = K^{s}/[H^{+}]$

Table II—continued
Adsorption Equilibria as Defined: (B) for for Hydrophobic Adsorption

B) Hydrophobic Sorption to Solid Phase containing Organic Carbon

$$A(aq) \rightleftharpoons A(s) \tag{1}$$

$$K_D = \frac{[A(s)]}{[A(aq)]} \quad \frac{[\text{mol kg}^{-1}]}{[\text{mol}\, \varepsilon^{-1}]} = \left[\frac{e}{kg}\right] \tag{2}$$

K_D is found to increase with the organic carbon content of the solid phase and with the hydrophobicity of the solute, K_{OW}.

The latter is expressed as the octanol-water partition coefficient

$$K_{OW} = [A_{oct}]/[A_{aq}] \tag{3}$$

Thus, the partition coefficient K_D (or K_P) can often be expressed as

$$K_D = b\ f_{OC}\ (K_{WO})^a \tag{4}$$

where f_{OC} = the (weight) fraction of organic carbon in the sorbent, a and b are constants; a is often around 0.8 (cf. Fig. 5).

Adsorption Controlling the Geochemical Fate of Pollutants and Reactive Elements

Adsorption influences the distribution of substances between the aqueous phase and particulate matter, which, in turn, affects their transport through the various reservoirs of the Earth. The affinity of the solutes to the surfaces regulates their (relative) residence time, their residual concentrations and their ultimate fate. Adsorption needs to be characterized in terms of the chemical and physical properties of water, the solute, and the sorbent. Two main basic processes in the reaction of solutes with natural surfaces are:

(1) the formation of coordinative bonds *(surface complex formation)*, and

(2) *hydrophobic* adsorption, driven by the incompatibility of the non-polar compounds with water (and not primarily by the attraction

of the compounds to the particulate surface) (Stumm, 1992). The adsorption by both processes can be quantified readily (Table 2).

Simple chemical models for the residence time of reactive elements and pollutants in oceans (Whitfield and Turner, 1987), lakes (Sigg, 1987), soil and groundwater systems (McCarty et al., 1981; Cederberg et al., 1985) are based on the partitioning of chemical species between the aqueous solution and the particle surface.

Figures 4a, 4b, 5, and 6 illustrate that adsorption to naturally occurring particles is very efficient, even at very low concentrations of solutes. Figure 5 illustrates that suspended particles, even at low concentrations, have a pronounced effect on the speciation of heavy metals and, in turn, on their reactivity and toxicity. The concentration of most pollutants (on a mol per kg or mol per liter basis) is much larger in the solid phase than in the solution phase. Figure 6 shows the sorption of non-polar organic hydrophobic substances to solid material that contains organic carbon. The sorption should be interpreted as absorption, i.e., a dissolution of the hydrophobic compound into the bulk of the organic material usually present as a component of the solid phase. The sorption may be compared by the partitioning of the solute between two solvents water and the organic phase. The distribution (or partition) coefficient, K_D, often also referred to as K^P.

$$K_D = \frac{\text{mol sorbate / mass of solid}}{\text{mil solute / volume of the solution}} \left[\frac{\text{e}}{\text{kg}} \right]$$

Figure 6 shows that the absorption increases with the content of the organic material in the solid phase and the hydrophobicity of the solute. Thus, K_0 is found to correlate with the organic carbon content and the hydrophobicity of the solute, K_{ow}, (Schwarzenbach and Westall, 1981) (see Eqs. 3 and 4 in section B of Table 2).

Adsorption in Groundwater and Soil Systems

Soil and aquifers can be looked at as giant chromatography columns. The concepts on transport of chemicals accompanied by concomitant adsorption and desorption has been borrowed from chromatography theory. There are, however, a few differences of importance (particles in soils and aquifers are polydisperse); one needs to distinguish between saturated and unsaturated subsurface zone; cracks and root zones may lead to preferential flow paths.

The theory for the effect of adsorption-desorption on homogeneous saturated media, using the one-dimensional form of the advection-dispersion equation, is established (e.g. Freeze and Cherry, 1979) and can be used for simple systems.

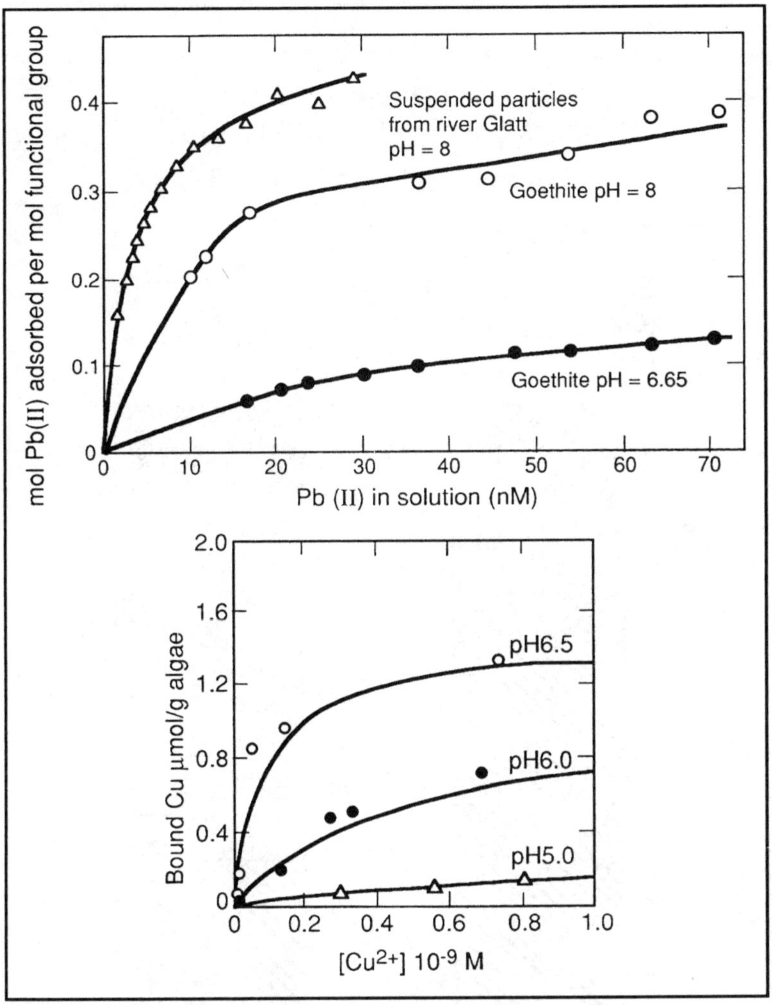

Figure 4a. Adsorption of heavy metal ions to the surface of goethite and natural particles. Surface complex formation of Pbz# on goethite and the surface of natural particles (Glatt River). The data are interpreted in terms of Langmuir adsorption isotherms.

Figure 4b . Cu(II) binding isotherms for algal surfaces at different pHs. Bound Cu was determined by AAS measurements of extracts of the algae after reaction with Cu at given pH; the reaction was carried out in a suspension containing 0.01 M KNO_3, 1.89 - $2x10^5$MNTA, 0.1-$1.8x10^{-5}$MTCu and 75 - 120 mg^{-1} algae dry wt. All isotherms follow a simple Langmuir equation at low coverage. The resulting binding constants and capacities are log K 8.4, 9.1 and 10, Fmax 9 10, 1.7 10^{-6} and 1.4 10^{-6} mol g for pH 5.0, 6.0 and 6.5, respectively (From Xue and Sigg, 1990).

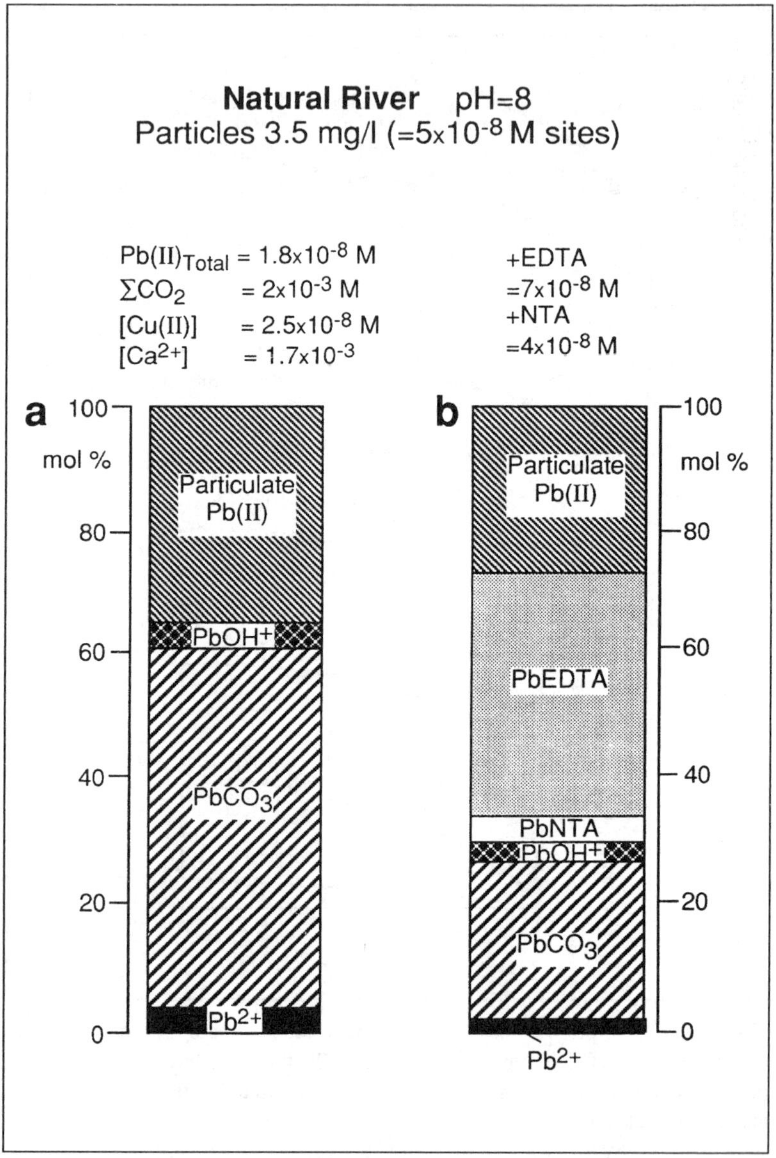

Figure 5 . Speciation of Pb(II) in Glatt River. The concentrations given for CO_2, Pb(II), Cu(II) and [Ca^{2}k] as well as for the pollutants EDTA and NTA are representative of concentrations encountered in this river. The speciation is calculated from the surface complex formation constants determined with the particles of the river and the stability constants of the hydroxo-, carbonato-, NTA- and EDTA-complexes. The presence of [Ca^{2+}] and [Cu^{2+}] is considered (From Miner and Sigg, 1990).

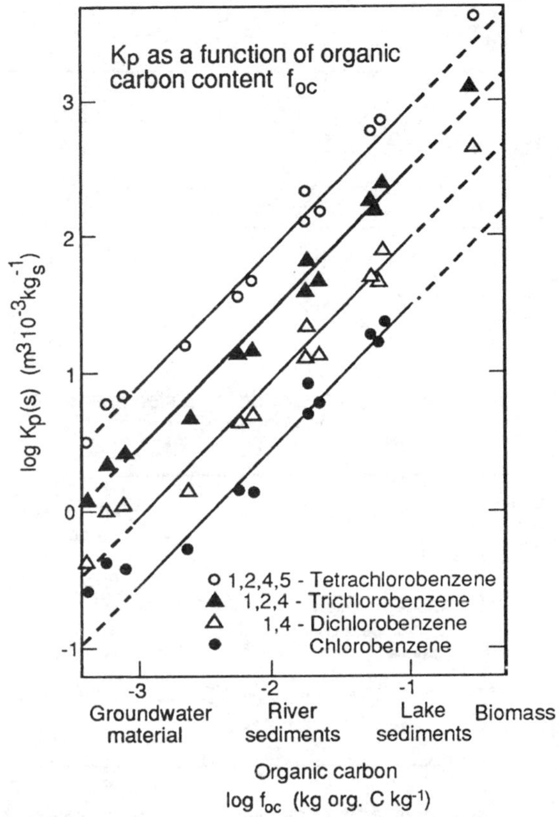

Figure 6 . The distribution of organic substances between water and representative solid materials of different organic carbon content, f^{oc} (Modified from Schwarzenbach and Westall, 1981).

In single cases (neglecting dispersion, linear sorption constant), the transportsorption equation can be written as:

$$-u\frac{\delta c_i}{\delta x} = \frac{\delta c_i}{\delta t}\left(1 + \frac{\rho}{\theta} K_P\right)$$

where:

u	=	linear velocity [cm s^{-1}]
c_i	=	concentration of species i
p	=	bulk density kg e^{-1}
θ	=	porosity (volume of voids/volume total)

(see Figs. 7a, b).

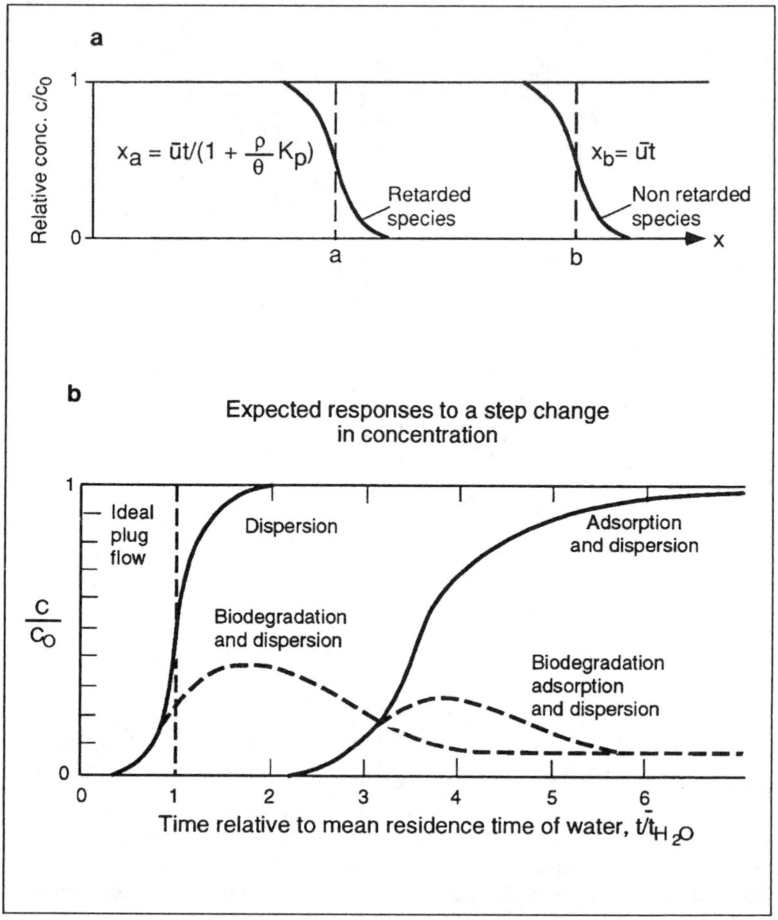

Figure 7a. Advances of adsorbed and non-adsorbed solutes through a column of porous materials. Partitioning of species between solid material and water is described by K_p. The relative velocity is given by:

$$\frac{\overline{u_i}}{\overline{u}} = 1/\left[1+(p/0)K_p\right]$$

Solute inputs are at concentration c_0 at time t > 0 (modified from Freeze and Cherry, 1979.) and are in linear velocity [cm s^{-1}] of groundwater and the retarded constituent, p = bulk density [kg e^{-1}], 0 = porosity, K_p = distribution or partition coefficient [e kg^{-1}], x = distance in aquifer, t = time.

Figure 7b. Illustration of the effects of dispersion, sorption, and biodecomposition on the time change in concentration of an organic compound at an aquifer observation well following the initiation of water injection into the aquifer at some distance away from the observation well. c represents the observed concentration and c_0 the concentrations in the injection water.

Although these theories have been tested successfully in column experiments, validation in real systems is still needed. In real systems, linear adsorption isotherms can often not be used. Furthermore, different solutes compete with each other for the available surface sites. Speciation and, in turn, distribution coefficients change progressively with distance (Cederberg et al.1985). It is essential that adsorption retards the transport of pollutants. Biodegradation in soil and aquifer systems is often better than in batch laboratory experiments.

Particles Regulate Water Composition of Reactive Elements in Lakes and Oceans

In the sea as well as in lakes, the ions of metals, metalloids, and other reactive elements interact competitively with soluble ligands (hydroxides, carbonates and organic solutes such as chelate complex formers including those formed by aquatic humus). Particle surfaces can tie up significant proportions of trace metals even in the presence of solute complex formers. The effects of particles on the regulation of metal ions are enhanced because the continuously settling particles (phytoplankton formed by photosynthesis and particles introduced by rivers) act like a conveyor belt in transporting reactive elements into the sediment.

Indeed, the partitioning of metals and other reactive elements between particles and water is the key parameter in establishing the residence time and, thus, in turn, the residual concentrations of these elements in the ocean and in lakes. The more reactive an element is, the more will it be bound to particles and the more rapidly will it be removed and the shorter will be its residence time:

$$\frac{C_{\text{part}}}{C_{\text{total}}} = \frac{\tau_P}{\tau_M}$$

where $C_{\text{part.}}$ and C_{total} are the particulate and total concentrations of an element, and τ_P and τ_M are the residence time of the particles and the residence time of this element with respect to removal via scavenging, respectively.

Phytoplankton. The surfaces of biogenic organic particles (algae) and organic biomass derived from them contain functional groups or ligands that are generally more efficient in binding bioreactive elements than inorganic surfaces, and are thus believed to represent in surface waters the most important scavenging phase and carrier from surface to deep water (see Fig. 4b).

In addition to adsorption processes, phytoplankton can absorb (assimilate) certain nutrient metal ions (or metal ions that are by the organisms mistaken as nutrients). As with other nutrients, this uptake can occur in stoichiometric proportions. The uptake (and subsequent release upon mineralization) of nutrients in stoichiometric proportions was claimed already in 1934 by Redfield. In referring to the atomic proportions C : N : P : Si, etc., one refers to the *Redfield Ratios.* This stoichiometry is well established (at least for the conventional nutrients) in oceanic waters; it has also been postulated for lakes.

Organic Matter in the settling material originates mostly from phytoplankton. The chemical composition can be compared to the Redfield stoichiometry of algae [$(CH_2O)_{706}$ $(NH_3)_{16}$ (H_3PO_4)]. In particles of Lake Constance (collected mostly during the summer), a mean composition $C_{113}N_{15}P_1$ was found, while the particles from Lake Zurich (collected over a whole year) had a mean ratio of C : P of 97 : 1 (N was not measured) (Sigg et al., 1987). In these lakes 15± 40 weight-% of the settling particles consists of organic matter. This fraction varies during the year due to seasonal variations of primary productivity.

Settling Biota as a Major Carrier of Heavy Metals. Among the components of settling material mentioned above, several lines of evidence point to the importance of biological material as a major carrier of trace metal ions. The binding of metal ions to surface ligands also represents a first step in the uptake of metal ions into the organisms (Fig. 8). Because organisms require a number of essential trace elements, such as Cu and Zn, it may be expected that these elements will be bound in certain ratios by algae similar to the Redfield ratio for C, N, P. If these elements are mostly bound to biological material in the settling particles, the ratios found in these particles should correspond to such a Redfield ratio.

For Lake Zurich and Lake Constance, the following tentative ratio was calculated:

$$(CH_2O)_{97}(NH_3)_{76}(H_3PO_4)_7Cu_{0{\cdot}006}Zn_{003}.$$

In Lake Zurich and Lake Constance, correlations between the contents of different trace elements in the settling particles and phosphorus, which may be used as an indicator of biological material, indicate that especially copper and zinc were likely to be associated with biological material (Fig. 9). In Lake Zurich, the highest sedimentation fluxes of Cu and Zn were observed during summer, simultaneously with the sediment fluxes of organic carbon and phosphorus (Sigg et al., 1987). Similar tendencies were also found for Cd and Pb, but the data were more scattered, due possibly to the tendency of these elements to adsorb to different types of particles.

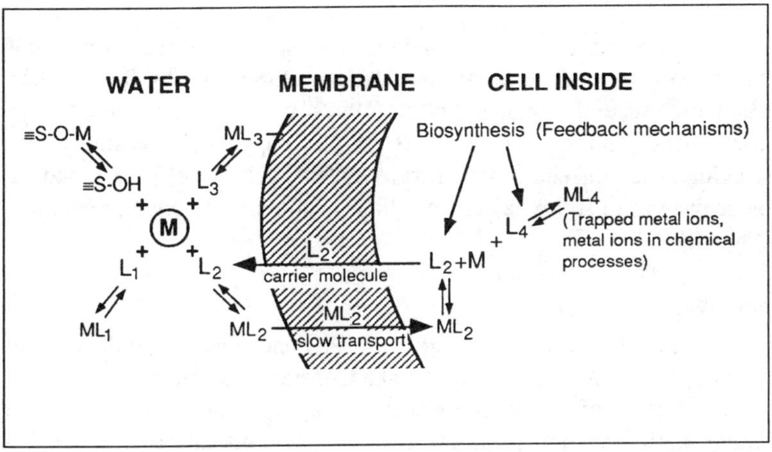

Figure 8. Uptake model in a simplified model, the metal ions equilibrate on the outside of the cell with biologically produced and excreted ligands L_2 or ligands on the cell surface L_3; these reactions are followed by a slow transport step to the inside of the cell. In the cell, the metal ions may be used in biochemical processes or become trapped in inactive forms as a detoxification mechanism. (From Sigg, 1987)

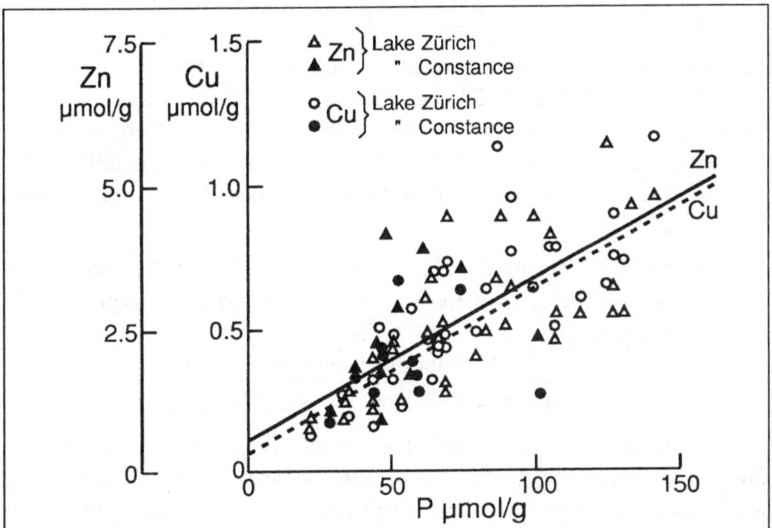

Figure 9. Concentrations of Cu and Zn as a function of P (phosphorus) in the settling material from Lake Zurich and Lake Constance. P serves as an indicator for the quantity of biological material present. The regression lines for Zn and Cu fall nearly together, indicating that rather constant ratios Cu : Zn : P are found in this material (From Sigg et al., 1987).

The idea to extend the Redfield ratio to some essential trace metals has been advanced by Morel and Hudson (1985) and by Sigg (1985). As shown recently by Morel et al. (1992), the case for colimitation of trace metals and major nutrients (C, N, P, S) appears increasingly convincing. The concentrations of trace metals such as Fe, Hi, Cu and Zn in seawater are so low as to limit their availability to aquatic microorganisms.

Oceans

In Figure 10, the depth profile of the Pb concentrations of the Central Pacific is compared with that of Lake Constance. In either case, the Pb concentrations of the surface waters are higher than in the deep water; atmospheric transport plays in both cases a significant role in supplying Pb to the surface water. The decrease in the concentration of Pb with depth occurs by particles that scavenge Pb(11) most efficiently. Patterson (e.g., Settle and Patterson, 1980) used data on the memory record of sediments to compare prehistoric and present-day eolian inputs. These data suggest that the present Pb(II) input is two orders of magnitude larger than that of prehistoric time.

As in lakes, other potential scavenging and metal regeneration cycles operate near the sediment-water interface. Subsequent to early epidiagenesis in the partially anoxic sediments, iron(II) and manganese(II) and other elements depending on redox conditions are released by diffusion from the sediments to the overlying water, where iron and manganese are oxidized to insoluble iron(III) and manganese(III, IV) oxides. These oxides are also important conveyors of heavy metals near the sediment surface.

Whitfield (1979) and Whitfield and Turner (1987) have shown that the elements in the ocean can be classified according to their oceanic residence times, τ_i:

$$\tau_i = \frac{\text{total number of moles of i in ocean}}{\text{rate of addition or removal (mol time}^{-1})}$$

which are, in turn, a measure of the intensity of their particle-water interaction. Thus, the elements that show the strongest interactions with the particulate phase have very short residence times; those elements that interact little with particles are characterized by long residence times (Fig. 11).

Concluding Remarks

How can we keep up with the steadily increasing pressures on our aquatic ecosystems and on the quality of our subsurface water reservoirs? Figure 12 gives an impression on the scale of length dimensions of importance in environmental and engineering science. While in a historic perspective, some of the natural sciences have developed from the large scales (cosmology) and other from small ones (electronic orbitals, molecular biology), environmental scientists and engineers had to occupy themselves more and more with entire ecosystems, regional cycling, and global hydrogeochemical cycles.

However, the reactions at the macroscale depend on understanding the processes at the microscale. Interfacial processes are important in most of the systems depicted metaphorically in Figure 12; often the processes at the surface may become rate-determining. Historically, impairment of water quality has occurred in three succeeding, but partly also overlapping phases (Clarke and Holling, 1984).

First Phase: Acute, Localized Pollution by Sewage and Industrial Wastes. This kind of pollution created unsanitary conditions in the receiving waters, imparted odor and taste, leading to the spreading of pathogenic organisms (water-borne diseases), created depletion of dissolved oxygen, favored saprobic indicator organisms, i.e., heterotrophic organisms responding to putrescible substances. Throughout the world, engineers have been very successful in ameliorating this kind of problem. This kind of localized pollution is largely amenable to technological control.

Second Phase: Pollution by Synthetic Chemicals. Cultural evolution has been faster than natural evolution. In industrialized nations, industrial activities have grown faster than human population, agricultural production has been intensified by application of fertilizers and pesticides and energy production has increased exponentially. Many of the synthetic chemicals which have been added to the biosphere within the last decades bear little resemblance to the natural products of the biosphere. Because they are not readily biodegradable, many of these chemicals survive long enough in the environment. Some of these substances, even if they exhibit no acute toxicity, may nevertheless impair the self-regulation of aquatic ecosystems and damage their life support function; others tend to become concentrated in organisms, some may become harmful to human health.

Third Generation Problem: Interference in Hydrogeochemical Cycles. Humans in their social and cultural evolution continue to be successful in diverting energy to the advancement of their own civilization. Receiving waters reflect not solely the activities within their drainage area, but also the impact of emissions carried over large distances through the atmosphere. The rapid changes that have been observed in the last decades in chemical and biological properties of

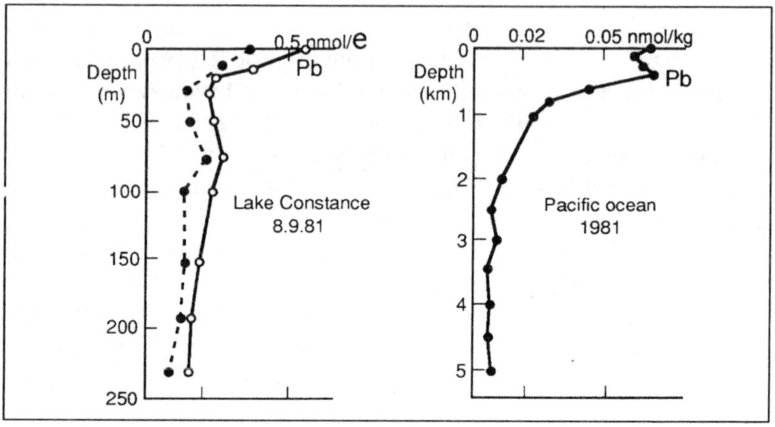

Figure 10. Lead profiles in Lake Constance (Summer 1981 data: Sigg, 1985) and in the Pacific Ocean (1981 data: Schaule and Patterson, 1981). The similar shape of these profiles, despite the difference in length scales (kilometers for the ocean and meters for the lake), illustrates the influence of the atmospheric deposition on the upper layers and the scavenging of Pb(II) by the settling particles (Modified from Sigg, 1985).

many coastal and fresh waters reflect the human influence on the environment.

As has been pointed out by Clarke and Holling (1984), "...we are moving beyond an age of acute localized, and relatively simple environmental problems reversible at economically measurable costs and politically realistic time and space scales. We are moving into a period of chronic, global, and extremely complex syndromes of ecological and economic interdependence". Despite intensive research, we understand only partially how chemical pollutants move between atmosphere, land, and water and what changes they undergo in their transport.

We have analyzed in this discussion, above all, adsorption processes and have discussed certain aspects of trace element uptake by phytoplankton. The processes considered here at the microlevel influence the major geochemical cycles. Understanding how geochemical cycles are coupled by particles and organisms may aid our understanding of global ecosystems, and low interacting systems may become disturbed by civilization.

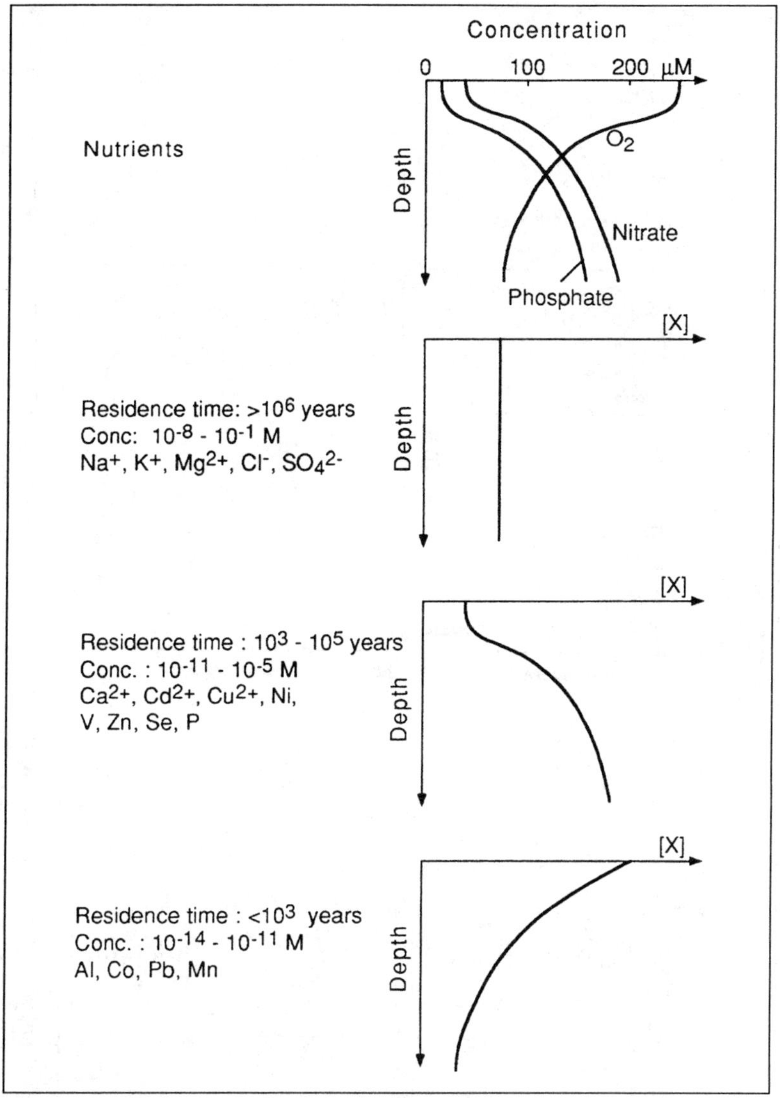

Figure 11. Schematic depth ocean profiles for elements. This figure is based on a classification of elements according to their oceanic profiles given by Whitfield and Turner (1987). Uptake of some of the elements, especially the recycled ones, occurs somewhat analogously to that of nutrients. There are some elements such as Cd that are non-essential but may be taken up (perhaps because they mimic essential elements) the same way as nutrients. The concentration ranges given show significant overlap, because the concentrations of the elements also depend on crustal abundance (Modified from Whitfield and Turner, 1987).

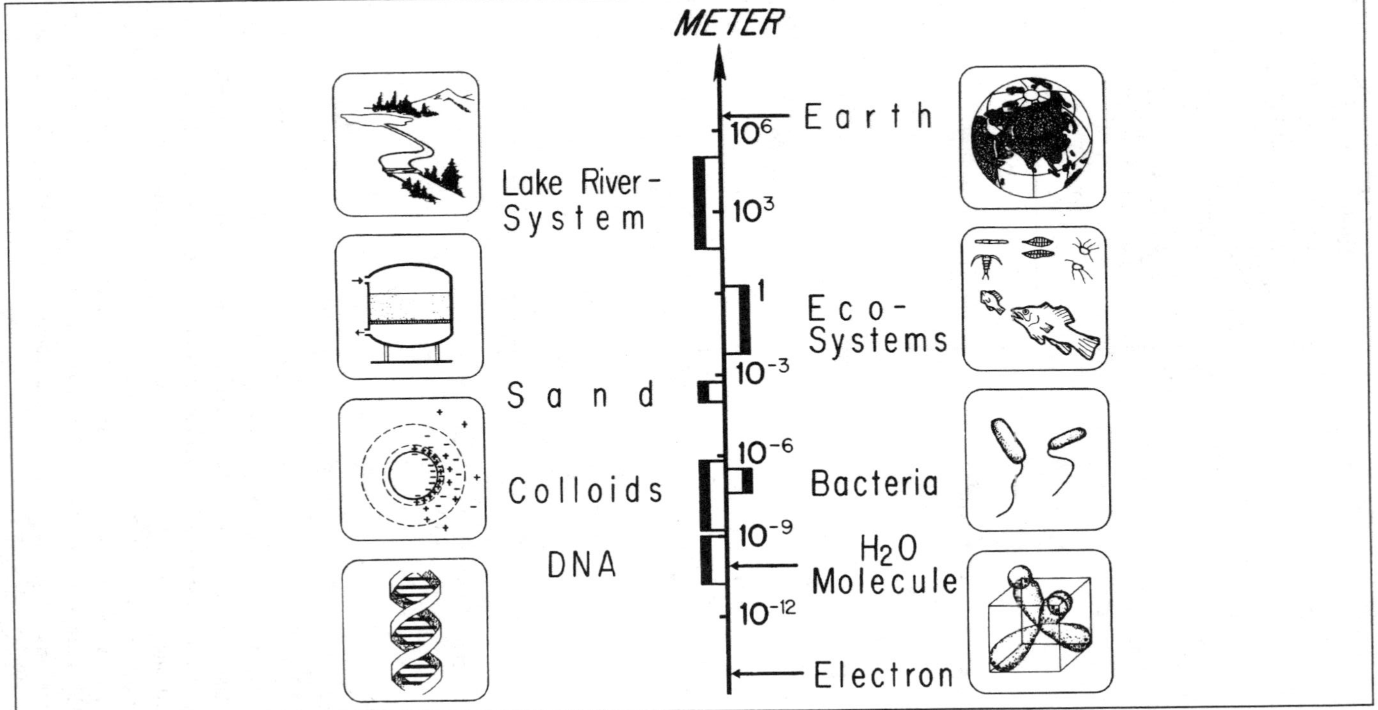

Figure 12. The length scales of importance for environmental engineers, scientists and hydrologists. Pollution is no longer a local problem; while we need to understand the processes at the microscale (colloid-water, organism-water, mineral-water interfaces), we need to consider these processes at the macroscale, in water technology, in entire groundwater and entire ecosystems and, finally, on the global cycling

REFERENCES

Buras, N. (1985), Foreword to 4th Kisiel Lecture.

Cederberg, G.A., R. L. Street, and J.O. Leckie (1985), "A Groundwater Mass Transport and Equilibrium Chemistry Model for Multicomponent Systems", *Water Resources Research 21*, 1095-1104.

Clarke, W. C., and C. S. Holling (1984), in T.F. Malone, and J. G. Roederer, Eds., Cambridge University Press, Cambridge, Massachusetts

Freeze, R. A., and J. A. Cherry (1979), *Groundwater*, Prentice Hall, NJ, 604 pp.

McCarty, P. L., M. Reinhard, and B. E. Rittmann (1981), "Trace Organics in Groundwater", *Environ. Sci. Technol. 15/1,* 40-51.

Morel, F. M. M., Hudson, R. J. M. and Price, N. M., (1992) Limitation of Productivity by Trace Metals in the Sea, *Limnol. Oceanogr.,* (in press)

Morel, F. M. M., and R. J. M. Hudson (1985), The Geobiological Cycle of Trace Elements in Aquatic Systems: Redfield Revisited, in W. Stumm, Ed., *Chemical Processes in Lakes*, John Wiley and Sons, New York, 251-281.

Müller, B., and L. Sigg (1990), "Interaction of Trace Metals with Natural Particle Surfaces: Comparison Between Adsorption Experiments and Field Measurements", *Aquatic Sciences 52/1,* 75-92.

Schaule, B. K., and C. C. Patterson (1981), "Lead Concentrations in the Northeast Pacific: Evidence for Global Anthropogenic Perturbations", *Earth Planet. Sc. Lett.* 54,97.

Schwarzenbach, R. P., and J. Westall (1981), "Transport of Nonpolar Organic Compounds from Surface Water to Groundwater, Laboratory Sorption Studies", *Environ. Sci. Technol.* 15,1360-1367.

Settle, D., and C. C Patterson (1980), Lead in Albacore: Guide to Lead Pollution in American, *Science* 207, 1167-1176.

Sigg, L. (1985), Metal Transfer Mechanisms in Lakes; The Role of Settling Particles", in W. Stumm, Ed., *Chemical Processes in Lakes*, Wiley Interscience, New York, 283-310.

Sigg, L. (1987), "Surface Chemical Aspects of the Distribution and Fate of Metal Ions in Lakes", in W. Stumm, Ed., *Aquatic Surface Chemistry*, John Wiley and Sons, New York, 319-349.

Sigg, L., M. Stumm, and D. Kistler (1987), "Vertical Transport of Heavy Metals by Settling Particles in Lake Zurich, *Limnol. Oceanogr.* 32,112-130.

Stumm, W. (1986), "Water, an Endangered Ecosystem", *Ambio 15(4),* 201-207.

Stumm, W. (1992) *Chemistry of the Solid-Water Interface Processes at the Mineral-Water and Particle-Water Interface in Natural Systems*, Wiley Interscience, New York.

Stumm, W., and J. J. Morgan (1970), *Aquatic Chemistry*, Wiley Interscience, New York, 583 pp.

Whitfield, M. (1979), The Mean Oceanic Residence Time (Mort) Concept. A Rationalization", *Marine Chemistry 8,* 101-123.

Whitfield, M., and D. R. Turner (1987), "The Role of Particles in Regulating the Composition of Natural Waters", in W. Stumm, Ed., *Aquatic Surface Chemistry,* Wiley Interscience, New York, 457-493.

Xue, H., and L. Sigg (1990), "Binding of Cu(II) to Algae in a Metal Buffer", *Water Resources Research, 24/9,* 1129-1136.

BIOGRAPHY

WERNER STUMM

He was born in 1924 in Switzerland. Werner Stumm obtained the Ph.D. degree in chemistry from the University of Zurich in 1952, and took additional graduate studies at Harvard University from 1954 to 1956.

From 1952 to 1956, Werner Stumm was a Research Fellow at the Swiss Federal Institute of Technology in Zurich. The following fourteen years he spent at the Harvard University, starting as Assistant Professor and culminating as Gordon McKay Professor of Applied Chemistry. Since 1970, Werner Stumm is Professor and Head of the Institute for Water Resources and Water Pollution associated with the Swiss Federal Institute of Technology in Zurich.

Werner Stumm has a rich list of publications. His classical *Aquatic Chemistry*, published by Wiley-Interscience in New York, was translated in several languages, including Japanese and Chinese. His *Chemistry of the Solid-Water Interface* was published in 1992. In addition, he edited a number of important monographs, among these the *Aquatic Chemical Kinetics* in 1990.

Werner Stumm earned a number of important honors. Starting with the Monsanto Prize of Pollution Control in 1976; he received honorary degrees from the University of Geneva (Switzerland) in 1987; the degree of D. Tech. (honorary) from the Royal Institute of Technology in Stockholm, Sweden, 1987; D.Sc. (honorary) University of Crete, Iraklion, Crete, Greece, 1988; D.Sc. (honorary), Northwestern University, Evanston, Illinois, 1989; D.Sc. (honorary), Technion – Israel Institute of Technology, Haifa, Israel, 1989.

In 1991, Werner Stumm became a Member of the US National Academy of Engineering.

FOREWORD

The central issue around which the activity called "Water Resources" revolves is the discrepancy—real or perceived—between water available as a resource and its use for a set of purposes. Both availability and utilization have the same attributes: quantities at given point in time, location (in three dimensional space), quality (physical, chemical, microbiological), and the institutional-legal framework within which water resources are developed and subsequently utilized. One means by which part of this discrepancy is overcome, especially in its quantity-temporal dimension, is by storing water.

Storage of water was practiced by mankind since the pharaohs in ancient Egypt. The dam Sadd-el-Kafara south of Cairo was built early in the third millennium BCE. Water, however, does not need to be stored exclusively in surface reservoirs. Aquifers, if properly managed, can be effective storage components in a regional water resources system. With proper tillage practices, moisture can be stored in the root zone of soils in sufficient quantities for crop production.

The matter of water storage, especially in surface reservoirs, is intimately connected with flow phenomena. To explain flow and storage, one has to rely on assistance from the science of fluid mechanics. The study of moving fluids presents difficulties, because one has to take into account the shape of the vessel containing the fluid and type of the conduit through which it moves. The complexity of the streamflow increases when considering movement of water in nature, from precipitation on a watershed to the runoff measured at its lowest point. The shape of the watershed, its geology and soils, the vegetative cover and various human interventions contribute to situations when water storage is either a blessed inspiration for socio-economic development, or a source of desperation produced by insufficient information and erroneous analysis.

Nathan Buras

WATER STORAGE: SOURCE OF INSPIRATION AND DESPERATION

VÍT KLEMEŠ

Expecting the Unexpected

When contemplating the topic for this lecture, I had before me the image of Chester Kisiel as I remember him when we last met - it was in 1972 - and as I will always remember him: listening attentively (I was telling him about my investigations of the Hurst phenomenon, a hot topic in those days), with a hint of a smile in his face, a co-conspirator's twinkle in his eye and a notebook (or was it a card?) in his hand, on a lookout for some interesting detail, idea worth noting, some subtle twist that might arouse his curiosity. In short, his expression conveyed a guarded but eager expectation of some unexpected intellectual stimulus. When it came, Chester's response was, as always, the same: "Have you written it up? Please, send me a copy". At the time I could not yet oblige; and when, two years later, I was ready[1], Chester was no longer with us. It is from such and similar memories that the topic for this memorial lecture has crystallized.

What I propose to talk about is not so much water storage but rather the UNEXPECTED which plays such a crucial role in our lives, puts excitement and frustration into our work and is responsible for the tangled pattern of the process known as scientific progress. Water storage enters into the picture only by default: most of my work has revolved around water storage, so it should not be unexpected that the specifics I have chosen to illustrate the substance of the UNEXPECTED revolve around it as well. And, borrowing Nick Matalas' observation made in a similar context in the first lecture of this series, I hope that this "substance, more so than the specifics,... would have been of interest to Chester" and perhaps also to you, especially if you have been attracted by the inscrutable manifestations of CHANCE.

Mind you, when I say chance, I really do mean CHANCE, i.e. the UNEXPECTED: the unforseen good luck and bad luck; the most obvious things missed and the unintended discoveries made; misrepresentations that have survived a century; wheels, even broken wheels, reinvented ... What I don't mean is statistics and probability theory with which chance is most commonly associated. After all, they cover only the most trivial aspects of it, namely those that **can** be expected - just think how these disciplines cling on to concepts like expected value, limit theorems, asymptotic laws, probable errors, maximum likelihood, stationarity, normality, linearity, etc. and how they avoid like a plague anything unexpected, improbable, unlikely, unpredictable, nonnormal, nonlinear, and the like! It is true that, as Nick Matalas remembered here eleven years ago, Chester Kisiel was attracted by these disciplines; but the irony of their (unavoidable) preoccupation with the EXPECTED did not escape him - he mocked it in one of his papers[2] in a prayer of "the theoretical hydrologist":

Oh, Lord, please, keep the world linear and Gaussian!

Remembering the perspicacious twinkle in his eye, I suspect probability and statistics might not have satiated Chester for long and he might have moved on to some more subtle aspects of the UNEXPECTED. I saw a hint of this in his intent to write a paper with Allan Freeze on quotations from famous individuals, as Allan revealed in the sixth lecture of this series. After all, such quotations are notable for the very reason that they typically contain some unexpected observation or idea.

How the World was Saved from Klemeš Storage Models

With some reluctance, I must admit that most of the credit for this must go to the late Leonid Brezhnev. For prior to his 1968 "fraternal help" to Czechoslovakia, which landed me in Canada, I found it far easier and more enjoyable to make my own discoveries and write about them in Czech than to read, in various poorly mastered foreign languages, about discoveries of others. Brezhnev changed all that. At the University of Toronto, I was expected to teach students the standard methods - the Rippl Diagram, the Puls Routing Method, the Moran Storage Model, etc. - and I used the opportunity to go to the original sources to read and learn about them. And the more I read the more fascinated I became by the gems I discovered, and also exasperated when I saw how often they were lost or misrepresented. Only then I realized what a deplorable practice it was to cite original sources from second-hand accounts, how wide spread this "science by citation" was and how often this cavalier attitude misplaced credit for original contributions and denied it to those who deserved it. I also realized how close I myself had come to perpetrating this reprehensible routine and resolved to make it the first rule of my work to give proper credit wherever it was due (even if it should go to Leonid Brezhnev).

From here, it was just a small step to see that most of my own discoveries were either trivial or redundant (as were, by the way, many of those made by others) and that it might be more prudent to search for the original gems buried in the literature than, by indulging one's own ego, risk possible future embarrassments.

My very first two papers illustrate the danger of embarrassment by triviality of what looked like a bright idea. In the mid 1950s, a colleague of mine, Jaroslav Urban, proposed a new graphical method for flood routing by a storage reservoir in his doctoral thesis[3]. It impressed me very much and, as I was playing with the Urban Method, I got two bright ideas how to simplify it. Both were duly published[4] and I was especially proud of the second one which found its way into some handbooks and textbooks as the Klemeš Method. However, when I set out to write a computer program for it some ten years later, it became obvious that my bright ideas were rather trivial as were, in fact, the differences between most of the graphical techniques: they completely dissolved in any numerical algorithm. The reason is simple since the whole problem boils down to solving, for successive intervals, a linear water balance equation (inflow minus outflow equals change in reservoir storage) with a nonlinear flow-rating curve of reservoir outlets (usually the outflow can be expressed as a power function of storage) - a problem for which every self-respecting 2nd-year engineering student can now write a program during a lunch break. The graphical procedures find the solution by fixing the position of the straight line representing the water balance equation in a given interval such that the outflow also satisfies the rating curve. Since the slope of this straight line is given by the scales of the plot (Fig. 1, right side) all that is needed is to identify its one point - and the choice of this point is the only difference between most of the graphical methods. Fig. 1 shows the points employed by some of the more and less famous of them, namely those of Puls (which is practically identical to the Russian Potapov's and the Swedish Ekdahl's methods), Sorensen, Urban and Klemeš. When I realized this in the late sixties, I expected that somebody would surely burst this bubble soon but, as years passed and nothing happened, I eventually decided to do it myself[5].

I think Chester would have appreciated the following twist of the story which also illustrates embarrassment via redundancy. A colleague from Manchester Institute of Technology once sent me a reprint of a paper describing a new graphical reservoir routing method. He explained that his retired colleague had asked him for some simple routing procedure and he directed him to my method. The gentleman used it and, while working with it, got an ingenious idea how to improve it. The improved method was described in the enclosed reprint[6] and, as I was amused to see, it was identical to the Urban Method which I had

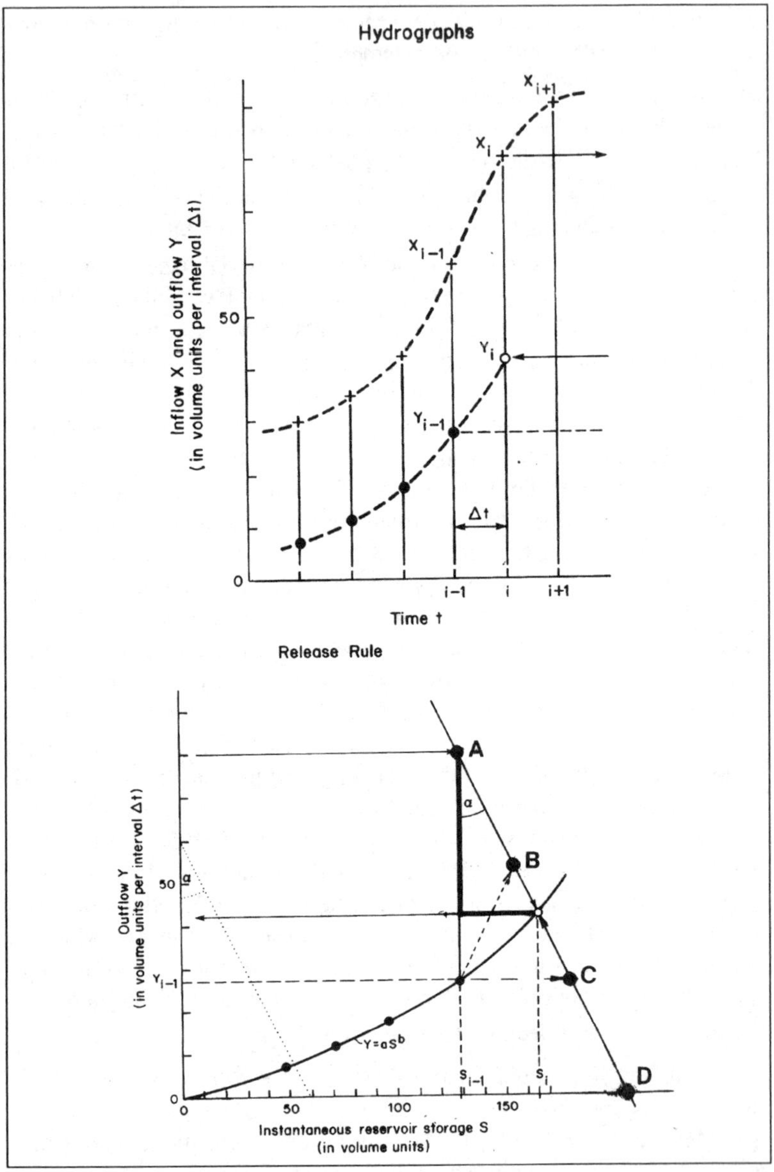

Figure 1. Common principle of reservoir routing methods: To find a position of the water-balance straight-line (whose angle α is given by the chosen volume scales) such that the difference between inflow and outflow in interval Δt is equal to the change in storage and the outflow obeys the release rule function.
One point is needed to fix this position. Examples of points employed: ***A*** *- Klemeš,* ***B*** *- Sorensen,* ***C*** *- Urban,* ***D*** *- Puls*[5].

been so proud to have simplified fifteen years before! When I revealed this to the author, he was amused as well and, in his reply, said it was not the first time that he reinvented the wheel (he did not realize it was a broken wheel in this case)- he had once derived a useful constant which he hoped would become known under his name, only to find out that it was already known in the fluid mechanics literature as the Froude Number.

In contrast, my next discovery was a breakthrough of great practical utility and deservedly is well known the world over. The only minor problem with it is that it is known as the Gould Storage Model - I barely finished the first draft of my paper when I found the whole thing neatly written up by Bernard Gould in an Australian engineering journal[7]. This shows that following the literature is an effective means for preventing embarrassment via redundancy. The unexpected twist came about fifteen years later when I gave a talk at the University of New South Wales. To honour Bernard, I talked about my applications of the Gould Model. After the lecture, Bernard told me in private that it was quite interesting but he would not recognize his model in what I was talking about and, as far as he was concerned, my talk was about a Klemeš Model. And so, for once, I may have missed a chance of not being redundant after all.

Fortunately, by that time my place in history had already been firmly secured by a discovery which earned me my doctorate. This discovery has proved to be completely immune to any danger of being found either trivial or redundant or, for that matter, of its scientific merit being questioned in any other way. It was a probabilistic method for the computation of the sub-annual component[8] to be added to over-year storage and it involved so many convolutions that there probably was, and ever will be, only one other person who has understood it. He included it in his book[9], now over quarter of a century old, but there is no doubt in my mind that neither of us remembers any more how the method really works and, being both retired, we are in little danger of being asked. The most unexpected aspect of this affair has been that this book, which cost less than one dollar when it was current, has recently appeared in an English translation[10] which sells for $165. This not only reflects favourably on the value of the Klemeš sub-annual storage model but further strengthens its immunity to potential criticism.

Could Wenzel Rippl Claim Damages from Harvard?

As far as I could find out, Wenzel Rippl owes his fame largely to the Harvard Water Program in whose publications his method was invariably used as the starting point for discussions of storage computations and a bench mark against which progress was measured. Older publications usually referred only to "mass curve" techniques but, starting with the profusion of references by Dorfman[11] to Rippl Method and Rippl Diagram in the 1962 classic volume by Maass et al.[12], these labels have become household words of the trade. But Rippl would have little reason to be grateful to the protagonists of the Harvard Water Program for his sudden fame.

Firstly, Dorfman's Rippl Diagram which found its way into most of the authoritative texts on the topic[13,14,15,16,17], was one that Wenzel Rippl had never used! While all these authorities represent it as a simple mass curve of reservoir inflow (and its tangents as mass curves of different rates of reservoir draft) as shown in Fig. 2a,b, the genuine Rippl's mass curve is an integral of the **differences** between the reservoir inflow and draft, i.e. a special case of the so called **residual mass curve** (Fig.2c).

Secondly, Thomas & Burden[18] set up Rippl's method as a straw man to be shot down because of its "defects" which, on closer examination, all come down to Rippl's failure to anticipate, in 1883, concepts advanced in the Harvard Water Program. His **method** of finding storage capacity from a time series of reservoir inflows was declared defective because he represented this time series by a historic streamflow record rather than synthetic flow sequences advocated by the authors. This alone makes one despair because the method has absolutely nothing to do with the nature of the time series to which it is applied! But what is even more unbelievable is the way his critics then set out to correct this "defect". They developed a **new procedure** for finding storage capacity for a given time series which they called the **sequent peak procedure** and which they proposed to be used "in tandem with synthetic hydrology". The point is that **the sequent peak procedure is identical to the original Rippl's method** as shown in Fig.2c! The only difference is that Rippl made the computations for his mass curve in a table by hand while his critics wrote a program to do the same by computer. In fairness to the innovators, one should not forget to mention two other improvements they introduced: they replaced Rippl's "crests and hollows" with "peaks and troughs" and changed his notation for the storage from J to S (as for their discovery that the computations must be run on two successive identical inflow series to get the correct value of storage capacity, this follows from the necessity to close the water balance over the computation cycle and has been common knowledge since the turn of the century[19]).

Thirdly, the explicit purpose of Rippl's paper[20] was to challenge the then common practice of computing the storage capacity of a reservoir

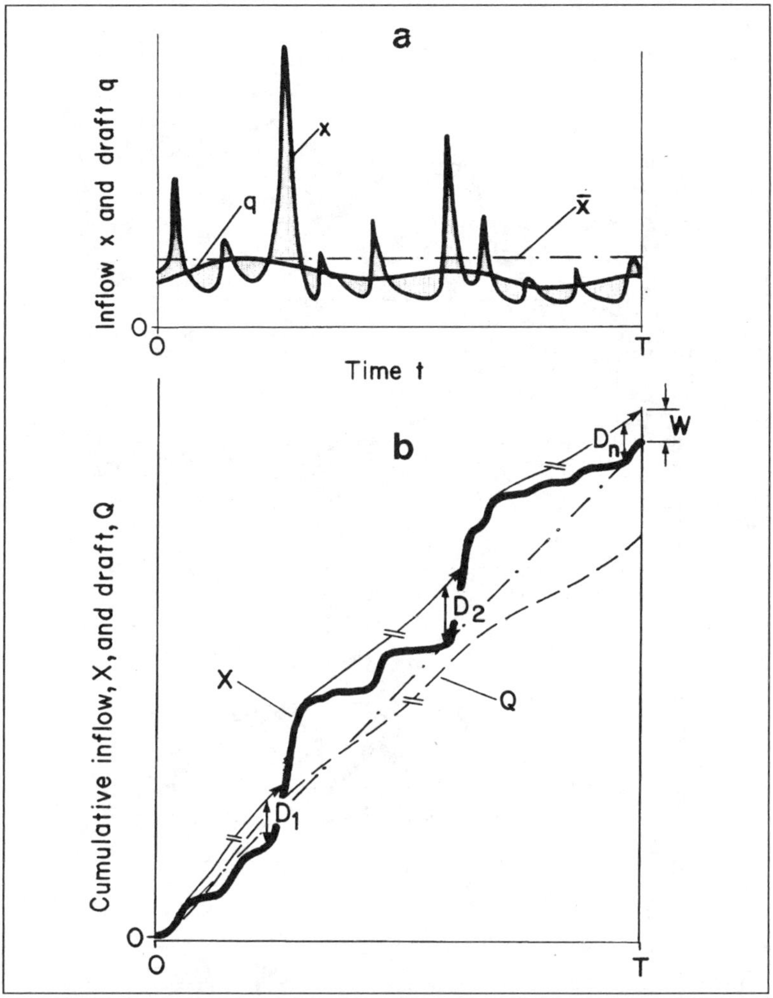

Figure 2. Definition sketch for various types of mass curves used for the determination of storage capacities (D_1, D_2, D_3) required in different periods to meet prescribed reservoir draft.

a - basic variables; **b** - common (absolute) mass curve; **c** - residual mass curve with respect to draft (Rippl's Method = sequent peak procedure); **d** - Hazen's procedure (reservoir behaviour diagram) is equivalent to Rippl's with deleted segments corresponding to periods of spillage (= periods when reservoir is full); **e** - common residual mass curve (computed with respect to mean inflow); its range R represents the storage capacity needed for "full regulation" (draft = mean inflow).

Note that for full regulation, Rippl's mass curve, Hazen's mass curve and the common residual mass curve are equivalent.

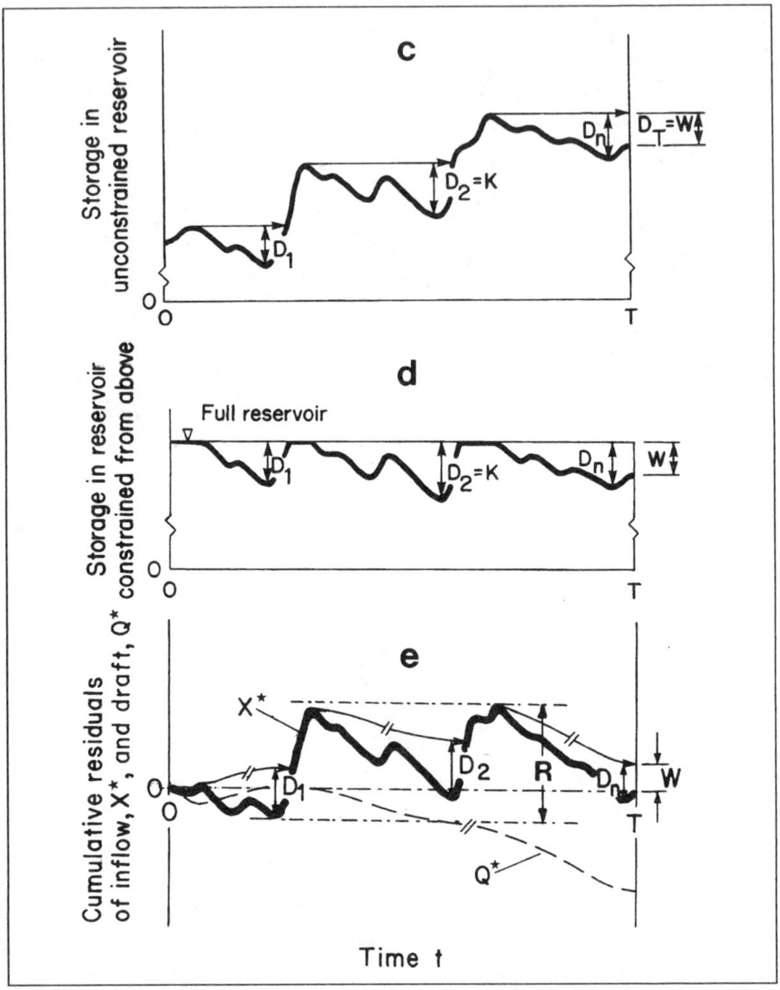

Figure 2—continued. Definition sketch for various types of mass curves used for the determination of storage capacities (D_1, D_2, D_3) required in different periods to meet prescribed reservoir draft.

a - basic variables; **b** - common (absolute) mass curve; **c** - residual mass curve with respect to draft (Rippl's Method = sequent peak procedure); **d** - Hazen's procedure (reservoir behaviour diagram) is equivalent to Rippl's with deleted segments corresponding to periods of spillage (= periods when reservoir is full); **e** - common residual mass curve (computed with respect to mean inflow); its range R represents the storage capacity needed for "full regulation" (draft = mean inflow).

Note that for full regulation, Rippl's mass curve, Hazen's mass curve and the common residual mass curve are equivalent.

as the water deficit in the single driest year of record. His main point was to show that a few only moderately dry years may require a larger storage capacity if they occur one after another. He made this clear in the very first sentence of the first section following the paper's Introduction: "The purpose of the storage-reservoir is to equalise the fluctuations of supply and demand during an **indefinitely long period of time**" and reemphasized it again at the end of his paper: "**The limitation of the time considered to a year is erroneous in principle, because the year is in reference to the question to be solved an unessential condition**" (emphasis added). His method was designed specifically to facilitate the determination of the long-term, over-year, storage requirements; to demonstrate its ability to accomplish this, he used an example in which the "critical period" extended over two years.

Given all these efforts to make his point clear, I wonder how Rippl would feel if he had a chance to read the following comment made - what an irony! - in his defense by one of his admirers, the late Mike Fiering[21]: "**In fairness to Rippl**, it should be pointed out **his technique was intended to investigate within-year storage fluctuations rather than over-year requirements**"(emphasis added). I remember how Mike himself felt when I once brought this to his attention and I doubt Rippl could have felt much worse. "Sometimes you pay a price when you take a shortcut and rely on judgements of those you hold in great esteem" he commented with a sigh. He didn't have to tell me more; I knew he had been "present at the Creation" [of the Harvard Water Program], as he later put it in the second lecture of this series (in which, by the way, he proudly made the point that it was being given in the year marking the one hundredth anniversary of the publication of Rippl's paper), and I had a fair idea about who the senior Creators pronouncing definitive judgements on Rippl might have been.

The American Debt to Allen Hazen

I know of no paper dealing with water storage that would contain a greater number and variety of original ideas than does the classic "Hazen (1914)"[22]. And I know of no other author whose so many ground breaking concepts have been neglected for so long in his own country.

Hazen's central idea was to introduce an explicit quantitative measure of hydrological uncertainty into the so called storage-yield function of a reservoir designed to control low flows. With such a measure (alternatively expressed as reliability or risk of failure), the function has the form shown in Fig. 3. Hazen's aim was to construct a function of this kind that would have a **general validity**. To do this, the uncertainty measure was to be based on **general patterns** of streamflow fluctuations expressed in a probabilistic form. This idea has never taken hold in America and, after Sudler's lonely attempt[23], Hazen's hope that, after his "only one step in the development...further study... will ultimately lead to more certain and accurate knowledge of the whole subject" would have been in vain had the "whole subject" not been taken up abroad. It was in the USSR where it inspired very vigorous studies starting in the 1930s and, over the next about forty years, produced (there as well as in other European countries) results of lasting value which I attempted to summarize a dozen years ago[24].

In the USA, only Hazen's idea to use **synthetic streamflow series** (which to him was merely a means for achieving the end result) was brought to fruition by Fiering in his doctoral dissertation[25] done under Harold A. Thomas. The Thomas-Fiering Model has since become a "Ford's Model T" of stochastic hydrology.

In pursuing his central idea, Hazen resorted to several ad hoc clever tricks in his paper which were meant merely to ease the burden of computational and drafting work he had to go through. He probably did not attach much importance to them and would not have expected that each of them would be enough to assure him of a permanent place in history.

Thus, for example, to simplify the plotting of the many storage distribution functions he had to analyze, he invented the Normal probability paper which has become a basic tool of statistics, an invention for which he is seldom given credit.

To plot the data points on the graph in some systematic way (and, as he put it, "with sufficient accuracy with a 10-in. slide rule"), he computed their positions on the probability axis by the formula $P= (2m-1)/2n$ (where m and n are the rank and sample size, respectively) which is still known as the Hazen plotting position in hydrology.

He also must have been one of the first to question the universal validity of the contemporary doctrine (based on Galton's fits of the Normal distribution to genetics data and Edgewoth's observation of the central limit theorem in the 1880s) that distributions of empirical data approach the "normal law of error" as the sample size increases. Distribution functions of his long flow and storage series ($n = 300$ and 402, respectively) showed a pronounced skew which made him doubt the accepted dogma: "Much more numerous data ... would be required to settle finally whether the law of error ... is strictly applicable to long-term records." Had he foreseen what difficulties the departures from

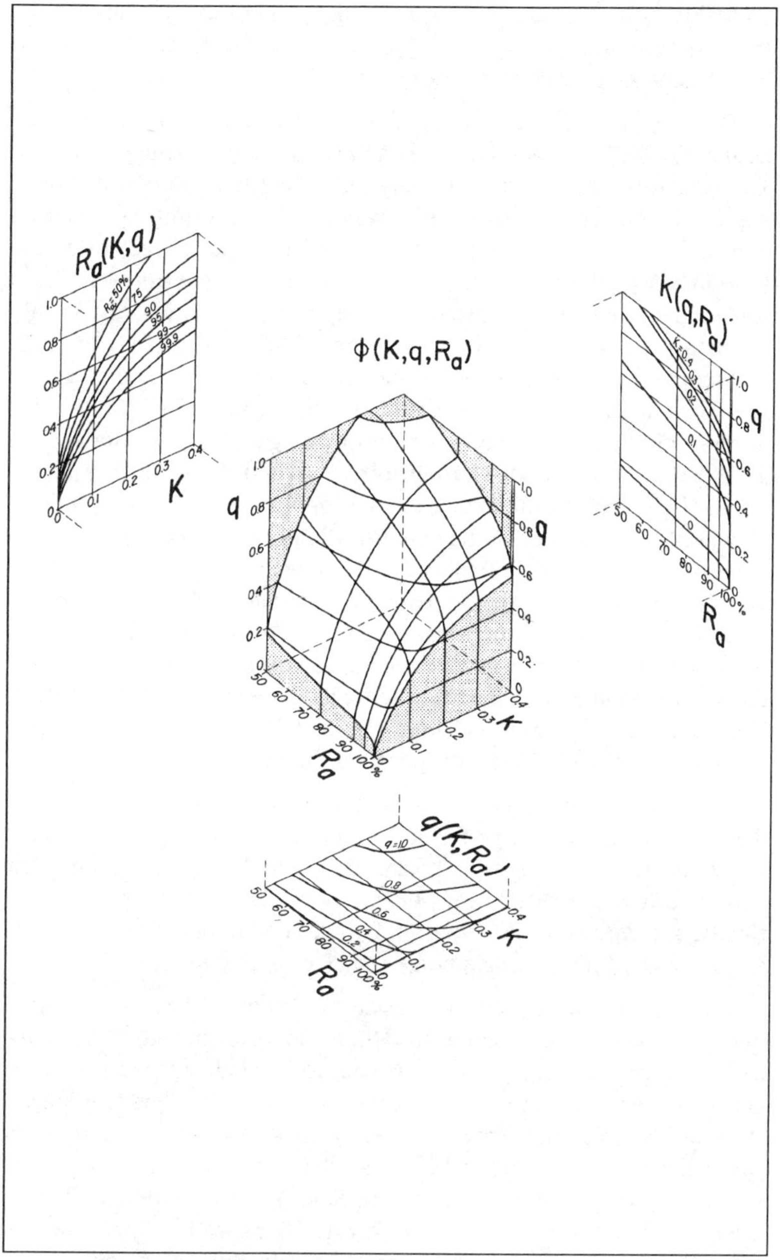

Figure 3. General reservoir storage-yield function $\phi(K,q,R_a)$. K - over-year storage (as fraction of mean annual runoff volume); q - draft (as fraction of mean flow); R_a - reliability (annual) = probability of non-failure year[24].

normality would cause to statistical and stochastic hydrologists, he might have turned a blind eye on his skew distribution curves and said the Chester Kisiel prayer himself.

When working manually with a 300-year long streamflow series as Hazen did, the application of Rippl's method becomes rather awkward. The reason is that, the draft usually being less than the mean inflow, the mass curve tends to run up and away from the drawing and, in addition, its irregular wavy shape makes it difficult to keep track of all the magnitudes of the individual storage values to be considered. Hazen solved this problem in an extremely simple but radical way: he simply dropped the segments of the curve corresponding to periods of spills (which are not needed anyway when one is only concerned with low flow regulation), thereby transforming it into a plot of the time series of reservoir storage fluctuations corresponding to the given draft. Such a plot has become known as **reservoir behaviour diagram**. It is shown in Fig. 2d where its difference from the Rippl scheme may seem minor. However, the enormity of the simplification it represents is obvious to anybody who has ever had to analyze (and understand!) in detail reservoir behaviour on a scale comparable to Hazen's. Moreover, Hazen's scheme substantially simplifies the computations and, unlike Rippl's procedure, is a genuine made-for-computer product. No wonder then that it has become the standard of the trade all over the world - except, it seems, the American academic circles where the original Rippl's method still reigns supreme disguised as the "sequent peak algorithm".

The irony of this situation can best be appreciated in the context of the now standard ritual, performed with a moving faithfulness in most American storage-related textbooks: first, Rippl's method is declared obsolete, then a paragraph of homage is paid to Hazen's genius and, finally, the sequent peak algorithm is presented in detail as the "modern technique" for solving storage reservoir problems.

Mike Fiering once told me that he had included Hazen's paper in the list of compulsory reading for all his graduate students. Alas, after becoming professors, his students seem to have abandoned this laudable practice of their professor. Recently, one of Mike's "grand graduate students", so to speak, (now a professor himself) sent me a draft of his paper in which, among other things, he advocated a new resiliency measure for reservoir performance - it was exactly the same measure Hazen used to characterize reservoir performance in his 1914 paper. The latter was not referenced and, it appeared, the author has never read it. However, since Klemeš was referenced several times for no good reason, I proposed a deal to the author offering him to trade two references to Klemeš for one to Hazen. That, so far, was all I could do for Allen Hazen in his native country.

Ups and Downs of the Residual Mass Curve

One discovery Allen Hazen did not make, but certainly would have liked to, was that of the "common" residual mass curve which integrates the deviations from the mean of a time series. That distinction went to Charles Sudler who probably didn't think much of his discovery himself. He used the curve to reduce the amount of drafting he had to do in his (otherwise rather abortive) attempt to advance Hazen's work on a general probabilistic storage-yield relationship (cf. 23).

This simple trick makes the residual mass curve a powerful and flexible tool. In storage analysis, it does away with the plotting of a separate mass curve for every different value of the draft. Instead, the same curve can be used for any draft because different drafts can be represented by tangents of different slopes (or curved shapes if they are non-constant) as shown in Fig. 2e; the added advantage is that the plot does not run up and away but unfolds neatly in the horizontal direction as do Hazen's "behaviour diagrams" which of course must be drawn separately for every different draft.

Hazen immediately saw the significance of Sudler's "incidental" innovation and commented[26]: "This paper contains a contribution of real importance ... The use of a mass curve, in which is shown only the accumulated surplus or deficiency as compared with the mean, permits a more convenient representation and accurate study of the data ... After having tried this method on an example the writer wonders that it was not done long ago."

This writer also wonders, namely why this technique, adopted the world over, has never been advocated in American textbooks which, as a rule, only casually mention the use of the basic mass curve (Fig. 2b), moreover, misrepresenting it as the "Rippl Diagram".

It was the English engineer Harold Edwin Hurst who elevated the common residual mass curve to a position of prominence reaching far beyond the context of storage reservoir computations. When he used Sudler's technique in his studies of storage needed for full regulation (one where draft is equal to the mean inflow) of the Nile River by the large Aswan Dam[27], he didn't have the slightest idea that his humble engineering analysis would lead to one of the most unexpected discoveries about the statistical behaviour of long records of empirical phenomena, ranging from streamflow to tree rings, wheat prices, annual catches of Canadian lynx and beyond - the famous **Hurst**

Phenomenon. It was a pure case of serendipity and it is now so well known that there is no need to dwell on it here (cf.1).

What is not so much appreciated is that it established the common residual mass curve as one of the principal tools of time series analysis. And it was through the interest in this analysis that this curve has found its way back to America (though not to reservoir analysis where it came from). There was a time - and it was Chester's time here in Tucson - when it seemed that almost everybody, from the brightest minds at the IBM down to every other American graduate student in stochastic hydrology, was working on some properties of the residual mass curve. It has been one of the greatest inspirations to which water storage studies have ever led.

However, it may also easily turn into a source of desperation because of its simple "iron rule" which says that what goes up must come down again and vice versa. In other words, the plot of every common residual mass curve must come back to zero from which it has started. The curve has two features which conspire to make many an analyst see patterns that do not exist in nature and are pure chimeras created by the mathematics of the curve. One feature is that, unless a sample is extremely skew, about half of its deviations are above the mean and half are below; the other is that, as Feller[28] has shown, the sign of the first deviation tends to fix the shape of the curve for a long period. And so it happens that plots of common residual mass curves tend to exhibit up and down swings with typical lengths between 1/3 and 1/2 of the sample size. Innocent as this feature may seem, it has "proved" cycles of wide ranging periods in hydrological and climatic data, not to mention climatic changes!

Here is an example: Williams[29], after analyzing a number of precipitation and streamflow records, concluded:"If cumulative deviations from the mean are computed for hydrologic data, continuous periods of 10 yr to 35 yr or more will be revealed in which hydrologic records are consistently below or above their means" (a part of Williams' plots is shown in Fig. 4). Given the fact that most of his records were between 60 and 70 years long, it could have hardly been otherwise. Had he used 100 year long records, their residual mass curves would show cycles 20 to 50 years long. Conclusions similar to Williams' have been reached, on exactly the same grounds, by several Russian authors.

The point is this: the trends would be real if the storage where the deviations from the mean flows, etc., have been accumulated were real – that is, if the plots were "true Rippl diagrams" in the sense that the means represented real outflows from real reservoirs. As a matter of fact,

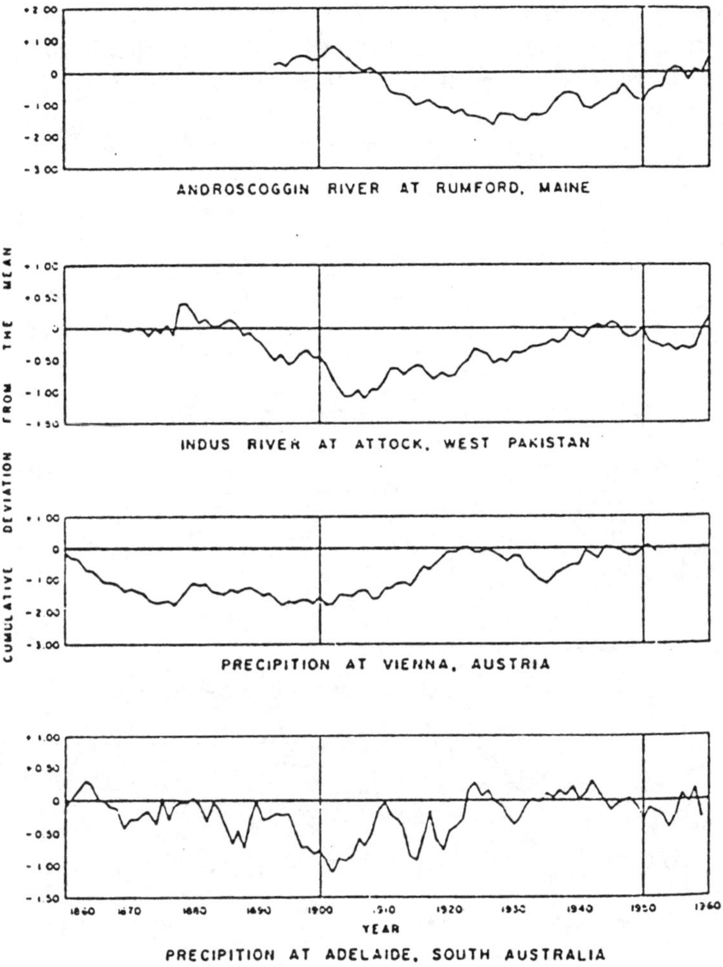

Figure 4. Some residual mass curves of hydrological records analyzed by Williams[29].

Nature does construct such true Rippl diagrams in the form of the historic records of fluctuations of glacier volumes, groundwater and lake levels! Unfortunately, or perhaps fortunately, **computer storage is not the same as water storage and can't cause climatic trends** as many people seem to believe. An example of the real thing and a computer chimera is shown in Fig. 5.

Even more misleading conclusions can easily be reached on the basis of residual mass curves of higher orders which, after about the 3rd

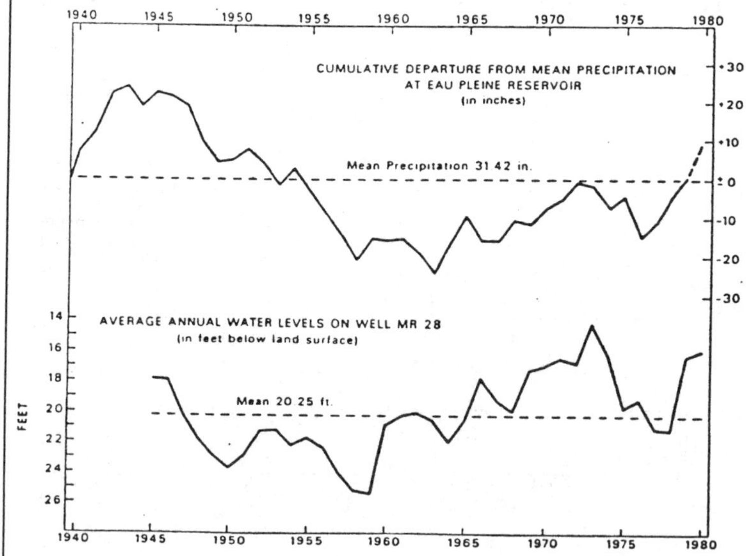

Figure 5. Observed annual groundwater levels and computed residual mass curve of precipitation at a nearby station.

order, converge to a pure sine wave with a period equal to the sample size, whatever the sample size may be and independently of the shape of the initial time series[30] (an example is shown in Fig.6). The most dangerous situation arises when a residual mass curve is computed from a historical record which itself already represents a "natural residual mass curve" of first or higher order. The result is then an almost perfect cycle extending over the whole historic record (cf. 30).

From Finite to Infinite Reservoirs

As every schoolboy knows, every water reservoir on earth is finite, from the smallest puddle, to ponds, lakes natural and man-made, to the world ocean. For dams, this is always certified in writing since every design specifies their maximum water level that must not be exceeded.

However, as Professor Moran explained in his classical 1959 monograph,

> "It being difficult to obtain explicit solutions for the finite dam, we attempt to simplify the problem. Since most of the difficulty arises from the boundary conditions..., it is natural to consider dams of infinite capacity and two cases now arise. We may consider what happens near the top of the dam when the probability distribution of [storage] is very unlikely to take values near zero. We may then regard the dam as infinitely deep ... Alternatively the conditions may be such that the probability of the dam ever being full is so small that we can take the dam as infinitely high and consider the probability distribution near the bottom."[31]

From that time on, many a mathematician writing on storage theory has started his paper with an apology to the engineer for the lack of realism in his forthcoming uninhibited musings about top-less or bottom-less reservoirs and assured him that the only reason for taking this liberty was mathematical convenience.

As an engineer, I have always (that is, since I first learned about these intriguing concepts while translating Moran's book into Czech) felt uneasy about such statements. I had a feeling that their apologetic tone was just a disguised discrimination, a subtle way of depriving the engineer of an equal right to share in this enviable source of inspiration. This inequity and injustice of being excluded from the privileged circles weighted heavily on me and I could not get the thing off my mind. But once, in a flash of insight, I realized that, in fact, **all real-life reservoirs, including river basins, were top-less!** Naturally, I was delighted by this discovery which assured the historically underprivi-

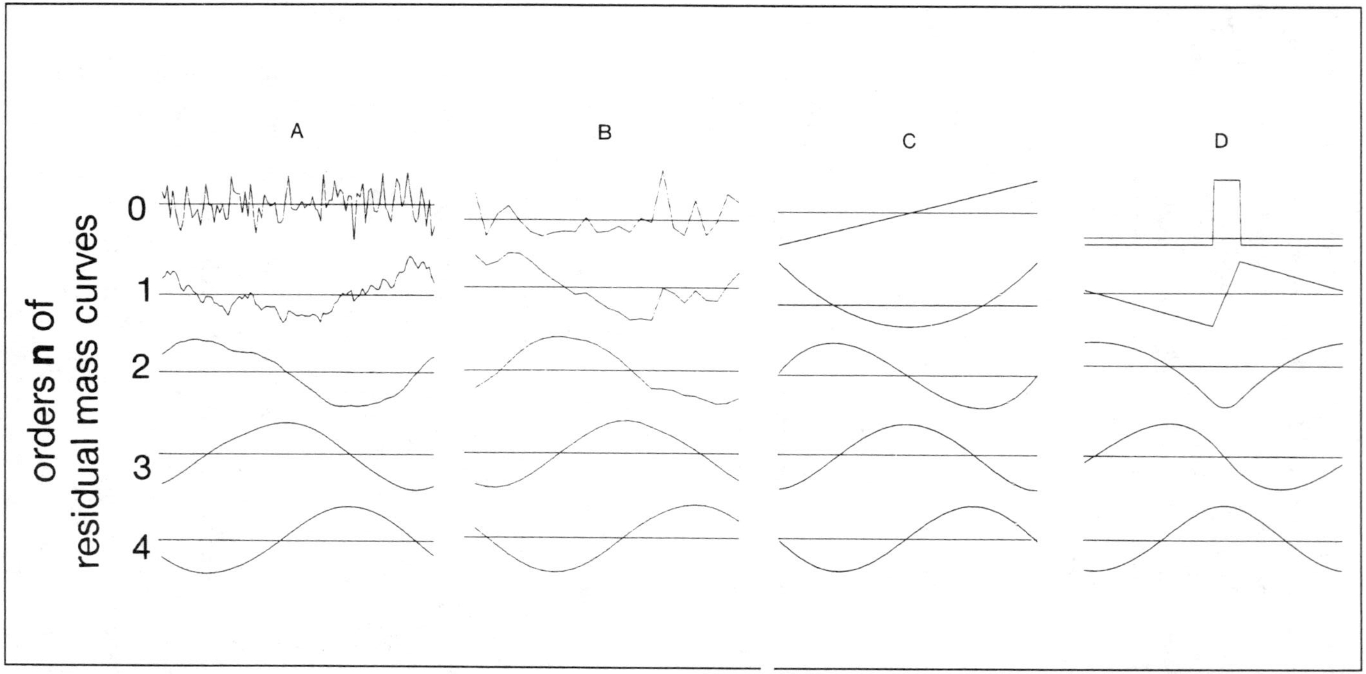

Figure 6. Examples of common residual mass curves of higher orders (order 0 represents the original time series). **A** - annual flows of the Rhine River in Basel, Switzerland (one hundred years, 1851-1950); **B** - random series generated from a lognormal distribution (sample size n = 25); **C** - linear trend line; **D** - rectangular puls[30].

leged engineers of equal opportunity and gained them access to the refined intellectual pleasures of the scientific elite. I raised the issue with Professor Moran at the earliest opportunity and assured him that there was no need for apologies: because precipitation distributions are always fitted with models that have infinite upper tails, no reservoir, be it a puddle or an ocean, can be prevented from exceeding its "maximum level" and thus from being effectively top-less. While admitting he has never thought about it in that way, Professor Moran, a kind man that he was, had no objections, especially when I pointed out that Noah's experience proved my point.

I consider the above finding my most important and lasting discovery and never since I made it ceased I to be fascinated by top-less reservoirs, to the point that I even bought one when the first opportunity presented itself. It was a beautiful small private lake near Ottawa, the Canadian capital, and it served as a source of inspiration to the whole family and many friends for a number of years.

An afterthought: Could it be that the reluctance of American professors to adopt Hazen's behaviour diagram stems from the fact that it so openly and clearly shows his reservoirs to be bottom-less? Or has their infatuation with the sequent peak algorithm something to do with the fact that it implies both bottom-less and top-less reservoir but makes neither feature too conspicuous?

Some Improbable Developments in the Probabilistic Theory of Reservoir Storage

It is paradoxical that it should be a probabilistic theory that is so richly endowed with improbable developments. It is true that, because of language barriers, eagerness to make one's own discoveries accompanied with reluctance to "waste time by reading all the obsolete old stuff" (as one young professor recently confided to me), some seemingly improbable occurrences should in fact be expected. Yet, the probabilistic theory of storage seems to be endowed with them over and above its rightful share.

Thus, for example, the equivalent of the famous 1954 Moran storage model for uncorrelated annual inflows was published by an obscure Russian engineer named Savarenskiy in 1940[32]. Ironically, not only was Savarenskiy never given credit for it in the West (it appears that the first English-language reference to him appeared in my paper[33] given at the same symposium where Chester Kisiel said the theoretical hydrologist's prayer), but had continuing difficulty getting it even in the USSR where his model was routinely attributed to Kritskiy &

Menkel, despite their repeated disclaimers. Another instance is the model for correlated annual inflows which was published in the same year, 1963, by Lloyd[34] in England and Kaczmarek[35] in Poland.

While all coincidences look unexpected, some are more unexpected than others. In this regard, the 1940 Kritskiy & Menkel[36] model for correlated seasons stands out in a class by itself (by the way, it was published in the same issue of the same journal as was the Savarenskiy model). When I once mentioned this model to Professor Moran, his face lit up with amused disbelief. He pulled out a paper from his shelf, leafed through it and made me read the following statement from it: "The above method neglects the possibility of correlation between flows in successive months. To take account of such correlation would be vastly more difficult." Then, with an obvious delight, he showed me the author and year of publication: Moran, 1955[37]. What apparently amused him most was the fact that Kritskiy and Menkel were just young engineers with no formal training in statistics and probability when they made their startling discovery, 15 years before he - a professor of probability and statistics - declared it vastly difficult.

However, as I subsequently found out, an even more unexpected aspect of their discovery was that Kritskiy and Menkel themselves apparently did not fully appreciate its extent. It took them nineteen more years to develop a rather complicated model for serially correlated (annual) inflows and they did not realize that a simple and elegant solution to the problem was implicit in their 1940 seasonal model: it just required to make the seasonal flow distributions and the season-to-season correlations identical and identify the "season" with the whole year. All that was needed to reformulate the theory was to drop the subscripts identifying the different seasons! Who knows whether the authors have ever realized this. When I did, Professor Kritskiy was already in his eighties and, though we still corresponded, I refrained from asking him.

This episode, together with my sincere interest in unhindered and rapid progress in storage reservoir research, has moved me to propose, at this point, the following extension of Chester Kisiel's prayer:

... and deliver us from all sources older than five years, including our own papers.

The Enigma of Negatively Skew Runoff

Prayers notwithstanding, one must face the facts of life. One of them is that the probabilistic distribution of annual runoff is skew. When I started working with storage reservoirs in the late 1950s, Hazen's doubts on this point had long been put to rest and a positive skew of annual runoff was taken for granted. The only unresolved problem seemed to be the mathematical form of the distribution. The favourite of the time and place was Pearson III with a zero lower bound, i.e. the 2-parameter Gamma. Its coefficient of skewness is double of the coefficient of variation, $C_s = 2\ C_v$. In the naivety of a young engineer, I thought there was some profound hydrological reason behind all this and, though I didn't know what it was, I was sure the authors of the textbooks advocating this preference knew. I was quite astonished when I later found they not only didn't know but mostly didn't even care(!) and were satisfied that the fit was good. This finding was more than a disappointment. It caused the first crack in my hitherto firm belief in the scientific nature of statistical hydrology, a crack that soon developed into a chasm dwarfing the Grand Canyon. But it inspired me to look for the hydrological basis of statistical properties of hydrological phenomena myself and the skew of annual runoff served as a good introduction into this fascinating area.

An incentive to start working on the problem presented itself with the appearance of Yevjevich's classic work[38] containing records of annual flows from 140 rivers around the world. However, when I plotted the C_s vs. C_v relationship for these data (Fig.7), my attention and curiosity were diverted to the unexpected fact that the skew coefficients were spilling over to negative values! By that time, thanks to Leonid Brezhnev, I already had access to the cream of western hydrologists and lost no opportunity to sound them out about the reason **why** that should be so. The results were quite devastating. In the best case, they had no idea, in the worst, they didn't even get my point. Thus, for example, the highest priest of contemporary American hydrology told me that I could "flip around" a positively skew distribution and get a good fit to a negatively skew one; his Australian counterpart suggested that, if I raised the C_s to the second power, I would "get rid" of my problem! I did not know how to make them see that I didn't want to fit anything, that I didn't want to get rid of the negative skew - that I just wanted to know what its cause might be.

One who got the point immediately was Chester Kisiel. When I discussed the problem with him in the summer of 1969, I already had some hints. I noticed, for instance, that the negative skew tended to be coupled with positive serial correlation in the flow series. That pointed to storage and, indeed, the rivers concerned had either large lakes, aquifers or glaciers in their basins. Chester listened attentively, scribbled in his notebook and then asked me to send him a copy of whatever

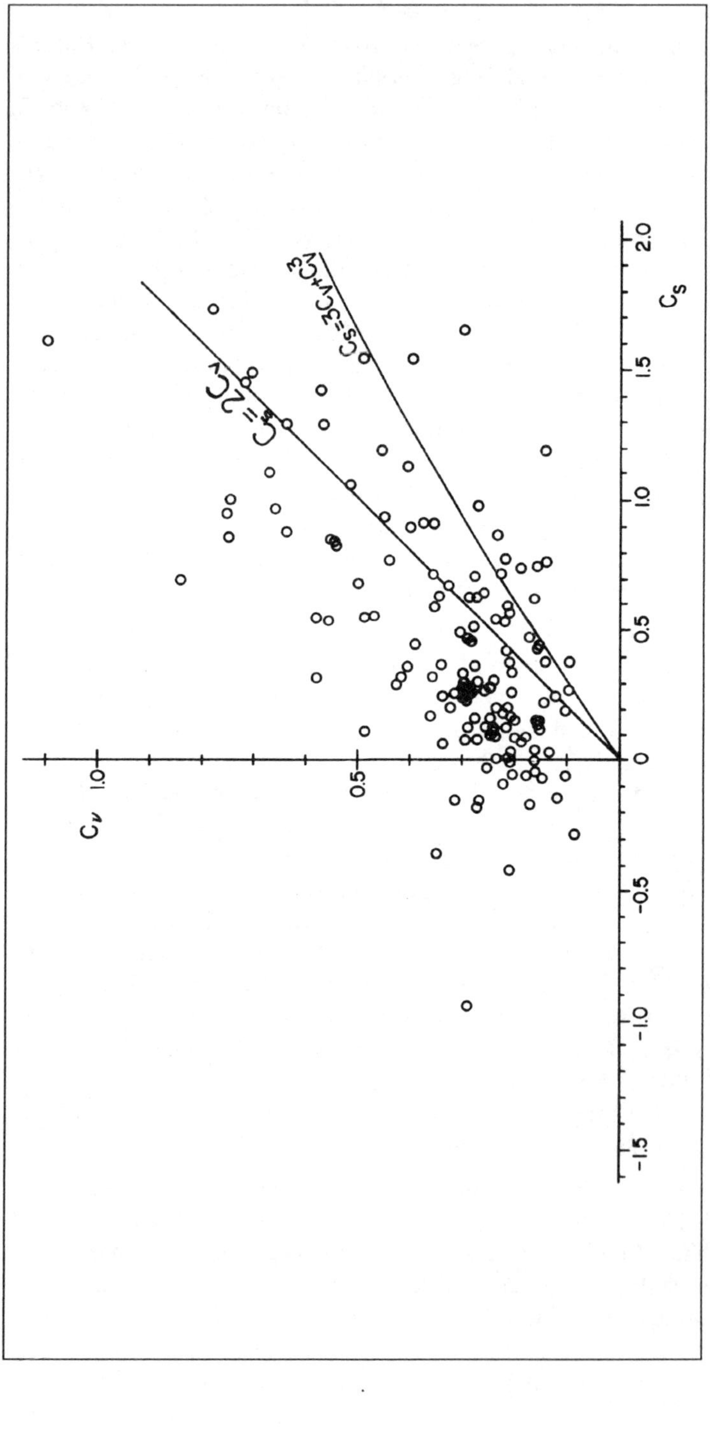

Figure 7. Relationship between the coefficients of variation (C_v) and skew (C_s) for historic records of annual flows for 140 rivers. The two lines show the functional relations for the Gamma-2 and the lognormal distributions.

I might write about it. I thought it was just a polite gesture and soon forgot about it. But, the next April I got a short note from Chester; it read: "Dear Vit: At last summer's meeting in Logan, Utah, we discussed your ideas on physical interpretation of skew coefficients. I wonder now if you ever had occasion to write this into a suitable format for discussion." So I send him what I had and, two months later, I read in his next letter: "Incidentally, I found your paper on skew coefficients quite interesting." I thought it was just a polite way of acknowledging the receipt of my paper. But Chester apparently meant it and kept thinking about the problem for, a year later, he returned to it again in one of his letters: "Your New Zealand paper[39] is of considerable value and interest to me. We need many more efforts along these lines."

Only then it dawned upon me where Chester's interest and persistence were coming from: It was from the deliberations that had led him to propose his prayer at the 1967 Fort Collins symposium. For, in his paper (cf. 2), the skew of output from hydrological systems was one of the problems he was puzzling about. He noted earlier results showing that "a single nonlinear reservoir, $S=KQ^n$, transforms a normal input to a non-normal skewed output with reduced variance ... in contrast to a linear storage system which reduces <u>both</u> variance and skewness of its inflow probability distribution" (Chester's emphasis). However, those results implied that zero variance and skew were the lower limits for hydrological variables. Chester summarized the situation in a cryptic comment: "Hydrologic systems are, in general, nonlinear in their transformation process. Very few theoretical results are available to predict output statistics." So, in retrospect, it was quite natural that Chester's interest was aroused by a physical mechanism that could lead to a negative skew of distributions of outputs from hydrological systems.

The last development Chester was to see along these lines appeared in a paper[40] which Professor Moran asked me to write for a symposium here in Tucson in September 1971, of which he was co-chairman. My paper showed clearly (and to my delight) that top-less nonlinear reservoirs could do the trick. I then pursued the idea further and Chester would have appreciated Fig. 8 in which I summarized my explorations some ten years later[41]. It shows that even a very positively skew input like the lognormal can be transformed into a negatively skew output by a suitably nonlinear reservoir, namely one whose outflow is proportional to storage raised to a power $b \ll 1$. But this is only a demonstration obtained by stochastic simulation - a general mathematical formulation proved to be beyond my reach and, as far as I know, Chester's cryptic comment that "very few theoretical results are avail-

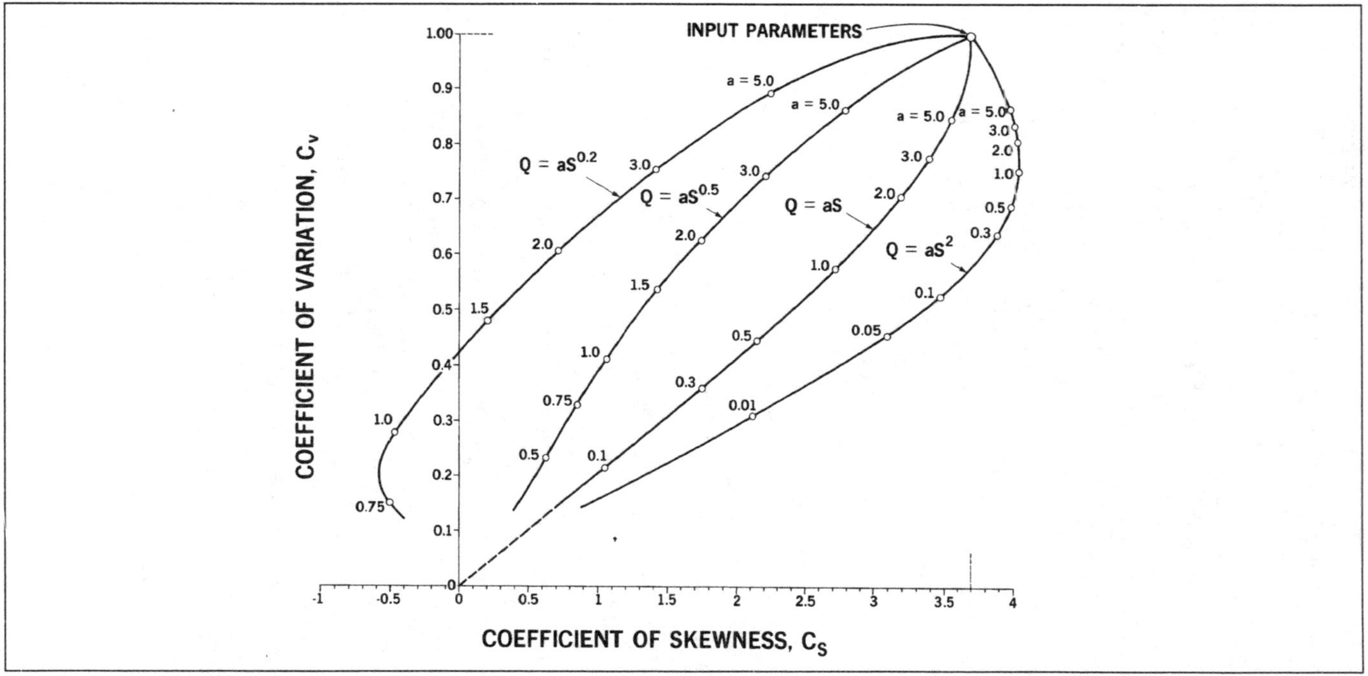

Figure 8. Relationships between the coefficients of variation (C_v) and skew (C_s) for output from one nonlinear reservoir fed with a random series generated from a lognormal distribution with C_v=1. The reservoir is defined by its release function $\mathbf{Q = aS^b}$, where **Q** is outflow rate, **S** is storage and **a,b** are constants. Each curve corresponds to a reservoir with a given value of **b** and a range of values **a**[41].

able to predict output statistics" (from nonlinear hydrologic systems) is still valid today, a quarter of a century later.

It has occurred to me that, already at the time of his request for the Tucson paper, Professor Moran must have known that my efforts (to get theoretical results for my top-less nonlinear reservoir) were doomed to failure but, a kind man that he was, he probably didn't want to discourage me by telling me so. For, in a paper on the future of stochastic modelling which he published shortly after[42], he wrote: "... it is clear that nonlinearity is an all pervading problem and here we are confronted, if not with a brick wall, at any rate with a hill of rapidly increasing slope." If it was a hill for him, no wonder it proved to be a Matterhorn for me.

In retrospect, I am glad I did not realize the futility of my "efforts along these lines" when I first set out to find out how things work in stochastic hydrology, which journey eventually led me to the high country of nonlinearity. If I did, I might have been fitting straight or "flipped around" Gamma-2 and other distributions to this very day and, who knows, may even have joined the distinguished gallery of experts who write authoritative treatises on the most rigorous scientific ways of doing it (currently, it is the method of linear moments, no doubt). But one thing is sure: I would have missed a lot of excitement (even frustrations can be exciting if they are of the right kind); a wealth of unforgettable intellectual exchanges with extraordinary people like Chester Kisiel, Pat Moran and Mike Fiering, to name just some of those whom I met on this journey and who have since passed away; and invaluable insights into how things work - and why they sometimes don't.

Let me close with a quotation which Chester would have certainly included in his intended paper, had he had an opportunity to write it. Its author is the late American physicist, Nobel prize winner, Richard Feynman[43]:

"The thing that doesn't fit is the thing that's the most interesting, the part that doesn't go according to what you expected."

REFERENCES CITED

1. Klemeš, V., 1974. The Hurst phenomenon - A puzzle? *Water Resources Research*, **10**(4), 675-688.

2. Kisiel, C.C., 1967. Transformation of deterministic and stochastic processes in hydrology. *Proc. Int. Hydrol. Symp.*,Vol. 1, 600-607, Fort Collins, Colorado.

3. cf. Urban, 1956, in [5].

4. cf. Klemeš, 1960a,b, in [5].

5. Klemeš, V., 1982. The essence of mathematical models of reservoir storage. *Canad. J. Civil Eng.*, **9**(4), 624-635.

6. cf. Cornish, 1974, in [5].

7. cf. Gould, 1961, in [24].

8. cf. Klemeš, 1963a,b, in [24].

9. Votruba, L. and Broža, V., 1966. *Hospodaření s vodou v nádržích.* SNTL, Prague.

10. Votruba, L. and Broža, V., 1989. *Water Management in Reservoirs.* Elsevier, New York.

11. cf. Ch. 3 in [12].

12. Maass, A. et al., 1962. *Design of Water-Resource Systems.* Harvard Univ. Press, Cambridge, MA.

13. Chow, V. T., 1964. Runoff. *Handbook of Applied Hydrology* (V.T.Chow, editor), Sec. 14, McGraw-Hill, New York.

14. Linsley, R. K. and Franzini, J. B., 1964. *Water-Resources Engineering*. McGraw-Hill, New York.

15. Hall, W. A. and Dracup, J.A., 1970. *Water Resources Systems Engineering.* McGraw-Hill, New York,

16. McMahon, T. A. and Mein, R. G., 1978. *Reservoir Capacity and Yield.* Elsevier, New York.

17. Loucks, D. P. et al., 1981. *Water Resources Systems Planning and Analysis*. Prentice Hall, London.

18. cf. Thomas & Burden, 1963, in [24].

19. cf. Stupecký, 1909, in [24].

20. cf. Rippl, 1883, in [24].

21. cf. Fiering, 1967, in [24].

22. cf. Hazen, 1914, in [24].

23. cf. Sudler, 1927, in [24].

24. Klemeš, V., 1981. *Applied Stochastic Theory of Storage in Evolution.* Adv. Hydrosc., **12**, 79-141.

25. Fiering, M. B, 1960. *Statistical Analysis of Streamflow Data.* (Doctoral dissertation), Harvard University, Cambridge, MA.

26. cf. Hazen, 1927, in Discussion to Sudler, 1927.

27. cf. Hurst, 1951, in [24].

28. cf. Feller, 1951, in [30].

29. cf. Williams, 1961, in [30].

30. Klemeš, V. and Klemeš, I., 1988. Cycles in Finite Samples and Cumulative Processes of Higher Orders. *Water Resources Research,* **24**(1), 93-104.

31. cf. Moran, 1959, in [24].

32. cf. Savarenskiy, 1940, in [24, 33].

33. Klemeš, V., 1967. Reliability Estimates for a Storage Reservoir with Seasonal Input. *Proc. Int. Hydrol. Symp.*, Vol. 1, 414-421, Fort Collins, Colorado. (In expanded form, J. Hydrol, **7**(2), 198-216, 1969).

34. cf. Lloyd, 1963, in [24].

35. cf. Kaczmarek, 1963, in [24].

36. cf. Kritskiy and Menkel, 1940, in [24].

37. cf. Moran, 1955, in [24].

38. Yevjevich, V. M., 1963. Fluctuations of Wet and Dry Years. Part I, *Hydrol. Pap.* No. 1, Colorado State Univ., Fort Collins, Colorado.

39. cf. Klemeš, 1970, in [40].

40. Klemeš, V., 1971. Some Problems in Pure and Applied Stochastic Hydrology. *Proc. Symp. Statist. Hydrol.*, 1-15, Tucson, Arizona.

41. Klemeš V., 1982. Stochastic Models of Rainfall-Runoff Relationship. *Statistical Analysis of Rainfall and Runoff,* 139-154, Water Resour. Publ., Littleton, Colorado.

42. cf. Moran, 1975, in [41].

43. Feynman, R., 1983. *The Pleasure of Finding Things Out.* Transcript of NOVA program #1002, WGBH Transcripts, Boston, MA.

BIOGRAPHY

Vit Klemeš

Vít Klemeš received a Civil Engineering degree and a Ph.D. in Hydrology and Water Resources in what used to be Czechoslovakia: the engineering degree from the Technical University in Brno (the Czech Republic) and the doctoral degree from the Technical University in Bratislava (Slovakia). For almost a decade, he worked in planning, design, construction and maintenance of dams and water resources systems with Czech and Slovak government agencies. Following this period, he joined the Hydrology and Hydraulics Institute of the Slovak Academy of Sciences where he did research in storage reservoir theory.

In 1966, Vít Klemeš came to Canada where he initially taught at the University of Toronto. During the period of 1972-1989, he did research in hydrology (primarily stochastic aspects of hydrological phenomena) and water resources systems at the National Hydrological Research Institute of the Canadian Federal Department of the Environment in Ottawa and Saskatoon; during the last ten years of the period, Vít Klemeš was the Senior Scientist.

While working for the Canadian Department of Environment, Vít Klemeš travelled broadly, giving lectures on all the five continents. He was a visiting professor at the California Institute of Technology, the Swiss Institute of Technology in Zurich (ETH), and at the Monash University of Australia. From 1987 to 1991, he was President of the International Association of Hydrological Sciences.

Since 1989, Vít Klemeš has lived in Victoria, British Columbia, Canada.